AUSTRALIAN SOIL
AND **LAND SURVEY**
FIELD HANDBOOK

FOURTH EDITION

The National Committee on Soil and Terrain

CSIRO

PUBLISHING

A catalogue record for this book is available from the National Library of Australia

ISBN: 9781486317660 (pbk)
ISBN: 9781486317677 (epdf)
ISBN: 9781486317684 (epub)

Published by:

CSIRO Publishing
36 Gardiner Road, Clayton VIC 3168
Private Bag 10, Clayton South VIC 3169
Australia

Telephone: +61 3 9545 8400
Email: publishing.sales@csiro.au
Website: www.publish.csiro.au
Sign up to our email alerts: publish.csiro.au/earlyalert

Cover art prepared by Helen Walter and Andrew Biggs

Cover design by Cath Pirret
Typeset by Envisage Information Technology
Printed in China by Leo Paper Products Ltd

CSIRO Publishing publishes and distributes scientific, technical and health science books, magazines and journals from Australia to a worldwide audience and conducts these activities autonomously from the research activities of the Commonwealth Scientific and Industrial Research Organisation (CSIRO). The views expressed in this publication are those of the author(s) and do not necessarily represent those of, and should not be attributed to, the publisher or CSIRO. The copyright owner shall not be liable for technical or other errors or omissions contained herein. The reader/user accepts all risks and responsibility for losses, damages, costs and other consequences resulting directly or indirectly from using this information.

CSIRO acknowledges the Traditional Owners of the lands that we live and work on across Australia and pays its respect to Elders past and present. CSIRO recognises that Aboriginal and Torres Strait Islander peoples have made and will continue to make extraordinary contributions to all aspects of Australian life including culture, economy and science. CSIRO is committed to reconciliation and demonstrating respect for Indigenous knowledge and science. The use of Western science in this publication should not be interpreted as diminishing the knowledge of plants, animals and environment from Indigenous ecological knowledge systems.

The paper this book is printed on is in accordance with the standards of the Forest Stewardship Council˚ and other controlled material. The FSC˚ promotes environmentally responsible, socially beneficial and economically viable management of the world's forests.

Aug24_01

CONTENTS

Contents

CITATION

If reference is made to the Field Handbook as a whole, give reference as follows:

- in text
  National Committee on Soil and Terrain (2024)

- in references
  National Committee on Soil and Terrain (2024) *Australian Soil and Land Survey Field Handbook*. 4th edn. CSIRO Publishing, Melbourne.

  If reference is made to a specific chapter (e.g. Landform), give reference as follows:

- in text
  Speight *et al.* (2024)

- in references
  Speight JG, Biggs AJW, Morand D (2024) Landform. In *Australian Soil and Land Survey Field Handbook*. 4th edn. CSIRO Publishing, Melbourne.

PREFACE TO THE FOURTH EDITION

The review undertaken for this edition was the most comprehensive since the publication of the original text. It was conducted as previous reviews have been, by a working group under the auspices of the National Committee on Soil and Terrain (NCST). Representatives were from states and territories, academia and the Australian Soil Classification Working Group.

The review was prompted by a number of factors, including changes in technology and methods, as well as recognition of inconsistencies with associated texts, such as *The Australian Soil Classification*. It is important that the latter is supported by the ability to describe attributes in a manner that enables classification. Another priority has been to expand the content of the Field Handbook to include terms relevant to a wider range of users and acknowledge the increasing frequency of disturbance of soils – for example, contaminated land and re-engineering of soil profiles in mining and agriculture.

The revision of the Field Handbook was also stimulated by the National Soil Judging Competition, which has been convened by Soil Science Australia since 2012. The competition has provided an important venue for knowledge sharing and discussion of soil and landscape description and classification. Importantly, it has challenged professionals and academics to evaluate more carefully the nuances of terminology, in the context of teaching new generations. The Field Handbook has always contained some necessary leeway in definitions, which existed for a number of valid reasons. However, such leeway can also be a source of unwanted variance in field description. Consequently, a key focus of this revision has been to remove ambiguity of interpretation wherever possible, but at the same time allow for the difference between a concept (as depicted in the Field Handbook) and reality in the field.

The Field Handbook has its origins in soil survey. While the amount of soil survey has declined in some states in recent years, the need to describe soils and landscapes in a consistent, standardised manner has increased, for an ever-widening range of purposes, not just soil survey.

CHANGES IN THE FOURTH EDITION

There have been many changes made in this edition, some obvious, some less so. Changes generally consist of:

- removal of redundant text
- re-ordering of text
- improvements to existing definitions
- addition of new terms.

There is increased cross-referencing to other standards that have developed over the last decade, in particular for vegetation and land use. Some older terminology that is no longer in common use has been removed from the Field Handbook, but these terms remain available for use via previous editions.

Terms used in *The Australian Soil Classification* that were not present in the Field Handbook have been added and some definitions refined within both texts – for example, the definition of a B horizon. The addition of new terms in this edition also recognises that the Field Handbook is not static – it must evolve as the science does. As with the original text, consideration of international methods has occurred. Some terms new to this edition have been obtained from other soil description texts, such as FAO (2006). Other terms introduced in this edition have been in common use in Australia for decades, but never previously formally defined. The full list of changes is described in a companion document (in press at the time of publication). Items removed from this edition are listed in Appendix 1.

Revision of the vegetation chapter was guided by a vegetation working group, coordinated by the Commonwealth Government Department of Climate Change, Energy, the Environment and Water (DCCEEW) and the Terrestrial Ecosystem Research Network (TERN). The working group constituted representatives from all state and territory governments. It evaluated how extensively the vegetation chapter of the third edition was used across the continent, observing that the second edition (Walker and Hopkins 1990) continued to be widely used and only a small component of the third edition had been adopted. Some sections have been amended from the third edition (not previously included in the second edition). Since the third edition was published, several Australian jurisdictions have created legislative frameworks for assessing vegetation condition. The vegetation chapter does not, however, provide a detailed discussion on the different approaches to defining condition. The third edition accommodated for non-native vegetation types, such as agricultural crops. The current chapter no longer differentiates between native and non-native vegetation, as the same observational principles of attributes can be applied.

Other changes implemented in the third edition vegetation chapter, including new height classes, an increased number of broad floristic groups, and different codes for some attributes, have not been retained in this edition, as they diverge considerably from the National Vegetation Information System (NVIS). The revised chapter provides guidance on how to generate a field description for a vegetation type at the site location, based on NVIS. The wetlands section now follows the Australian National Aquatic Ecosystem (ANAE) classification framework and provides guidance on the field attributes that are required. Additional attributes are also incorporated, including basal area, plant traits and phenology that could be useful to record in the field.

Some of the methodological changes predicted during the writing of the previous edition of the Field Handbook have not come to fruition and there is growing recognition that there is no substitute for good quality data collected in the field by trained experts, working to defined standards. As has always been the case, the primary purpose of this Field Handbook concerns terminology for the description of soil and land attributes (landform, land surface, vegetation). It can be used by a wide audience undertaking many activities and provides a basis for standardised communication of terms in Australia – whether that be verbally, pictorially or electronically.

It is hoped that the revision of this edition will set a new standard for the future, widening the user base for the Field Handbook, and in doing so, continue the valuable legacy of the original authors.

ACKNOWLEDGEMENTS

The Australian Soil and Land Survey Field Handbook Working Group was comprised of:

- Andrew Biggs (Chair, Queensland, Department of Resources)
- Jon Burgess and Pat Burley (Northern Territory, Department of Environment, Parks and Water Security)
- Stephen Cattle (Academia, University of Sydney)
- Brian Hughes (South Australia, Primary Industries and Regions SA)
- Darren Kidd (Tasmania, Department of Natural Resources and Environment Tasmania)
- Dave Morand (New South Wales, Department of Climate Change, Energy, the Environment and Water)
- Tim Overheu (Western Australia, Department of Primary Industries and Regional Development)
- David Rees (Victoria, formerly Department of Jobs, Precincts and Resources)
- Noel Schoknecht (Australian Soil Classification Working Group)

The vegetation chapter Working Group comprised the chapter authors and the following:

- Angela Muscatello (Victoria, Department of Energy, Environment and Climate Action)
- Anne Buchan (Victoria, Department of Energy, Environment and Climate Action)
- Belinda Allison (Commonwealth, Department of Climate Change, Energy, the Environment and Water)
- Brett Howland (Australian Capital Territory, Environment, Planning and Sustainable Development)
- Martin Mutendeudzi (Commonwealth, Department of Agriculture, Fisheries and Forestry)
- Robert Priddle (Commonwealth, Department of Climate Change, Energy, the Environment and Water)
- Ron Avery (New South Wales, Department of Climate Change, Energy, the Environment and Water)
- Wes Davidson (Commonwealth, Department of Climate Change, Energy, the Environment and Water)
- Illustrations by Bronwyn Bean

Others who have provided expert advice, support or ideas included:

- Lauren Eyre, Dan Smith, Mark Sugars and Nev Christianos (Queensland, Department of Resources)
- Henry Smolinski (Western Australia, Department of Primary Industries and Regional Development)
- Rob Moreton (Tasmania, Department of Natural Resources and Environment Tasmania)
- Bernie Powell (Australian Soil Classification Working Group)
- Ben Harms (Queensland, Department of Environment and Science; Australian Soil Classification Working Group)

- Land resources staff of the Queensland Government, who acted as an initial discussion forum for many review topics.
- Soil Science Australia

Jason Hill (former Chair, NCST), Darren Kidd (current Chair, NCST) and Peter Wilson (CSIRO) in particular are thanked for their governance, support and encouragement of the Working Group.

PURPOSE AND USE OF THE FIELD HANDBOOK

JG Speight, RF Isbell and AJW Biggs

PURPOSE

This Field Handbook is intended to contribute to the systematic recording of field observations of soil, vegetation, landform and land attributes in Australia. It attempts to:

- list attributes[1] thought necessary to adequately describe the nature of the landscape, in particular the soil and vegetation
- define these attributes whenever possible in a manner consistent with their use elsewhere in the world, but giving particular emphasis to Australian conditions
- define terms and categories for landform, vegetation, land surface, soil and substrate material that are based explicitly on the specified attributes
- provide codings for the various attributes, terms and categories to enable concise recording in the field and databases.

The linkage from field description to digital storage of data has long been important and the Field Handbook has, and will, continue to play a key role as a reference for natural resource databases and subsequent use of data. Observations made in the field may be supplemented by data from remote sensing, maps, records, laboratory analyses, experiments, local information etc.

This Field Handbook was prepared to meet the needs of surveys somewhat diverse in nature. Importantly, the process of observation at a point is not necessarily constrained to the process of mapping – that is, in the modern context, the term 'survey' implies the observation of attributes at a point in space and time, not whether the observation was part of a mapping exercise. Observations may be for a single purpose or form part of integrated activities, including soil or vegetation mapping, and biophysical, ecological, archaeological or other surveys, whether for agricultural, recreational, industrial, residential or other purposes, such as a general scientific

---

1 No distinction is made between the word 'attribute' and the word 'property'. Both mean 'characteristic' or 'trait'. 'Attribute' includes 'variable'. Observations produce *values* of attributes or properties.

inventory. The observations proposed are relevant to either point observations or mapping. Both are constrained by concepts of scale. Variability within the observational space must be considered within the context of the scale and accuracy of observational methods and the purpose of the work.

The recording of attributes of the site and adjacent landforms has two distinct purposes. Firstly, they may be directly relevant to land use or ecological function. Secondly, they are a link between the hidden physical and chemical properties of the landscape (for which data will always be scarce) and the visible properties of landform, surface material and vegetation that may be more readily mapped and catalogued.

Site attributes can link to other attributes both within a site and beyond it. On the one hand, they are intended to be correlated with soil and other subsurface properties observed at the site in an effort to discover significant relationships between them. On the other hand, site data are intended to establish local 'ground truth' values for the landform, surface material and vegetative properties that contribute to the more extensively developed characteristic image, 'signature', or pattern on aerial imagery or other remote-sensing record. Establishing short-distance variation for certain attributes is often very important in the selection of detailed study sites – for example, reducing internal variability in field trials or other research sites.

A key purpose of the Field Handbook is to provide a common and precise understanding of the meaning of terms used in all forms of formal and informal communication – from field notes to general discussion and publications. The Field Handbook is not a methods document *per se*, although methodology is implicit as part of defining many attributes and how they are recorded. The availability of methods documents for specific attributes/topics is growing and will continue to grow into the future.

USE

The Field Handbook is to be used as part of a systematic observational process in a landscape. The systematic logic starts with metadata concerning the site (who, when, where, what, how), progresses through landscape attributes, down through the soil profile to the substrate. The structure of the book reflects this logic. There is by no means a requirement to describe every feature at a site or in a landscape, but it is widely acknowledged that failure to record critical attributes can lead to a significant reduction in the value of data. Attributes are measured or estimated in as quantitative a manner as possible and users should always be cognisant of the accuracy and precision of field measurements. Given field observation relies upon individual skill, there is always a degree of subjectivity involved. All efforts should be made to minimise this through appropriate training and methods, but it must be recognised that accurate recording of some field attributes requires significant expertise.

For many attributes, there is a suggested scheme of classes provided in this Field Handbook, but this does not preclude the observation and recording of actual values where appropriate. Provision has not been made for all conceivable attributes, rather the focus is on those most commonly observed in Australia, in the context of the landscapes encountered in this country. This does not preclude the description of other attributes. The Field Handbook should, when necessary, be used with companion texts that provide more detail about specific topics or attributes – for example, methods of vegetation, gully or acid sulfate soil description. Relevant metadata concerning methods should always be recorded.

It is important that above- and below-ground attributes are described as they are, and not as they may or may not have been. It is also important that landscapes and profiles be described as factually as practicable, but genetic inferences are inevitable. Where these are used, the basis on which the inference is made should be noted so the user is aware of assumptions made.

The observational process in the field must deal with the varying dimensions of different entities in the landscape (soil profile, landform, vegetation community etc.) and the differing ways in which these may be described (point, plot, transect, defined or ill-defined space). The Field Handbook caters for these variations but at its core is the concept of an observation made within a site space (see page 5). The concept and dimension of a *site* may vary slightly with survey purpose. In the case of soil description, it is centred on a *soil profile*. For vegetation survey it may represent a *plot* or transect.

There is a clear distinction between *observation* (the purpose of the Field Handbook) and *classification*, which relies upon observed properties. There are many possible formal and informal classification systems that may be applied to soil, vegetation and landform attributes. Classification is not dealt with in the Field Handbook, although it is acknowledged that it is a core use of data collected in the field.

Throughout the Field Handbook, suggested codes for each attribute described appear in red. All dimensions are expressed in SI units. For some attributes, the codes 0 or Z are used to explicitly record the absence or apparent absence of a feature. The use of such codes is encouraged for all features observed, as the absence of a record can lead to issues with subsequent data use and interpretation.

THE SITE CONCEPT

JG Speight, RC McDonald and AJW Biggs

A *site* is a small area of land considered to be representative of the *landform, vegetation, land surface, soil* and other land features associated with the point of observation. A distinction is made between a *site* (being a dimensional and conceptual entity) and an *observation*, being a record of one or more attributes at a specific point in space and time. At times, the terms are used interchangeably, but they each have a specific context and should be used appropriately.

One or more attributes may be observed within the site space at one or more points in time. Some attributes, such as *substrate lithology* and *landform pattern*, are unlikely to change over time and are likely to extend beyond the site space. However, there is typically an intent to minimise variability within the site space to a specific instance of key attributes – for example, the observation is constrained to one *landform element*, within one landform pattern and geology. Other attributes, such as *surface condition*, *vegetation* and *microrelief* may be constrained very locally within the site space and/or vary with time – either naturally, or as a result of anthropogenic influence.

The extent of a site is variable depending on the purpose of the observation, the method of observation and the attributes of concern. Despite this, certain dimensions are generally appropriate and should be consistently applied unless there are clear reasons for departure. The nature of the site space should be explicitly recorded.

Landform attributes may be consistently observed, irrespective of type, purpose or method of survey. Observe landform element attributes over a dimension of approximately 20 m in radius (1257 m²) and landform pattern attributes over a dimension of approximately 300 m in radius (28.3 ha). Refer to the Landform chapter (page 13) for more detail.

The attributes *erosion, aggradation, microrelief* and *inundation* are observed across the same 20 m radius used for *landform element* attributes. Observe other *land surface* attributes within a dimension of approximately 10 m in radius (314 m²), such as *slope, aspect, disturbance, surface coarse fragments, surface condition, rock outcrop* and *runoff*. Given these attributes can be spatially variable in short distances, care must be taken to ensure the point of observation is representative.

A few *land surface* attributes refer simply to the point of soil observation, namely *elevation, drainage height* and *depth to free water*.

In some instances, an observation may be representative only of a space smaller than 10 m in radius. For example, in some *gilgai*, the vegetation, land surface attributes and soil all differ between the mound and depression, over distances <10 m. In such instances, the extent of the space that comprises the observation for those features is dictated by the extent of the microrelief component.

At times, a *plot* (a fixed sample space) may be employed to represent a site. This is particularly common in vegetation survey but also in some types of detailed soil or experimental studies. The shape and dimension of a plot should always be determined in the context of variability evident in attributes at the site. Variability within a plot should be minimised as much as possible but if it is unavoidable, its nature should be recorded.

GENERAL

RC McDonald, RF Isbell and AJW Biggs

Apart from location, there are other core data that must be recorded for any site, regardless of the attributes being observed. These include who created the observation, when and the purpose of the observation. It is a given that other pertinent metadata will be recorded as required in any field studies – for example, metadata concerning methods.

DESCRIBED BY

Give the first three letters of the surname and one initial, for example:

NORK for K.H. Northcote.

DATE

Give the date the observation was made, for example 23 December 2021, as *23122021*.

SITE PURPOSE

There may be one or more purposes for making an observation at a site. The purpose(s) of a site should be recorded. This allows for improved utility of data within a single collection (database). The codes below are a provisional list, but by no means exhaustive. A key reference site is one in which a comprehensive description of all attributes is undertaken – for example for soils, morphology, chemistry and physical attributes are described as recommended in McKenzie *et al.* (1995). This differs from a soil/land attribute monitoring site in which one or a few attributes are monitored, but the full suite of attributes are not characterised, such as a soil water monitoring site or a groundcover monitoring site.

A	*Acid sulfate soil survey*
C	*Ecological study site*
E	*Erosion survey site*
G	*Agronomic sampling*
K	*Key reference site*

M	*Soil/land attribute monitoring site*
S	*Soil survey site*
V	*Vegetation survey site*
Z	*Salinity survey site*

Sites may also be categorised as per Table 14.2 in McKenzie *et al.* (2008) or other schemes.

PLOT CHARACTERISTICS

If a plot is used at a site, there are key characteristics that must be recorded, including the dimension and shape of the plot. Observers should be very cognisant of variability of attributes within plots – use of fixed dimension plots necessarily creates a need for greater care in site selection to avoid undue variability within the plot.

Plot shape

Record the plot shape.

C	*Circle*
S	*Square/rectangle*
T	*Triangle*
O	*Other*

Plot dimensions

Record the dimensions of the plot in metres.

Plot shape	Dimension
Circle	*Radius*
Square	*Side*
Rectangle	*Length and width*
Triangle	*All sides*

Reference location

A reference location should be recorded for a plot (e.g. centroid or south-west corner) using a standardised approach, such as defined in some existing methods – for example, Muir *et al.* (2011). When appropriate, a compass bearing of the plot axis should also be recorded.

Plot shape	Reference location
Circle	*Centre*
Square	*Corner (e.g. the most south-westerly)*
Rectangle	*Corner (e.g. the most south-westerly)*
Triangle	*Apex*

LOCATION

LJ Gregory, RC McDonald, RF Isbell and AJW Biggs

There are two key components to the description of the location of a site/observation: the method by which the site location was chosen and the method by which the coordinate location was acquired. The method of selection of the site location was historically inferred via description of the *survey type* (described in the General chapter in previous editions of the Field Handbook). However, it is more appropriate that this is explicitly captured with the location, thus the concept has been moved into and expanded upon in this chapter.

It is likely, in nearly all cases today, that a coordinate location will be derived using a Global Navigation Satellite System (GNSS). Consequently, a significant amount of text present in previous editions of the Field Handbook has been removed from this chapter.

METHOD OF SELECTION OF SITE LOCATION

The most common practice for selection of a point in space to make an observation may be regarded as a *free* survey approach, which involves an element of expert knowledge and explicit or implicit site selection criteria – for example, the characterisation of a catena (McKenzie and Grundy 2008). *Grid* surveys are often used for very detailed (large scale) surveys. *Erosion* and *salinity* surveys are a specific type of site selection method, in which the location of the site is determined by the presence/absence of the feature being observed. *Statistical* methods for selection of site locations have become common in recent years. The *miscellaneous* method encapsulates site locations determined by ill-defined criteria or when the location of the site was determined by externalities such as land use/access – for example, it is only possible to observe the corner of a paddock due to cropping constraints. Transects are often associated with ecological surveys and linear corridor studies, such as road, rail and pipeline.

Record the method used to select the site location.

F	*Free survey*
G	*Grid survey*
M	*Miscellaneous*
P	*Purposive sampling for a specific context (e.g. erosion or salinity surveys)*

S   *Statistical method*

T   *Transect (linear corridor)*

METHOD OF ACQUISITION OF LOCATION

Record the method used to acquire the coordinate location of the site/observation.

I   *Geographic information system*

G   *Global navigation satellite system (GNSS)*

R   *Map reference*

S   *Survey*

When appropriate, record the GNSS survey method used to obtain the coordinates.

A   *Averaging GNSS*

D   *Differential GNSS*

S   *Single unit GNSS*

COORDINATES

Coordinates may be recorded in degrees or metric units. In all cases, it is important to record the datum and when appropriate, the zone. Although sub-metre accuracy of coordinates is typically not required in soil and land survey, it may be appropriate for some purposes, such as monitoring sites. Understand the limitations of the equipment used and the various factors that will affect the accuracy of the location. Sub-metre accuracy is usually only obtained through the use of differential techniques. Autonomous (single unit) methods can currently obtain <5 m accuracy under optimal conditions.

Datum

Record the datum of the coordinates. Older maps will generally be based on the Australian Geodetic Datum of 1966 or 1984 (AGD66, AGD84), while current maps should be based on the Geocentric Datum of Australia (GDA94 or GDA2020). If you are obtaining coordinates from a GNSS, the native datum is the World Geodetic System (WGS84). However, this may not be the display default, so check the settings. For further information, see the *Geocentric datum of Australia technical manual* (Intergovernmental Committee on Surveying and Mapping 2002).

WGS84   *World Geodetic System 1984*

GDA2020   *Geocentric Datum of Australia 2020*

GDA94   *Geocentric Datum of Australia 1994*

AGD84   *Australian Geodetic Datum 1984*

AGD66   *Australian Geodetic Datum 1966*

Projection

Coordinates may be projected onto the Universal Transverse Mercator (UTM) system. In Australia, this is called the Australian Map Grid (AMG) or Map Grid of Australia (MGA), depending on the datum used (AGD66/84 *vs* GDA94/2020 respectively). Do not use the Universal Grid Reference notation.

State whether the coordinates are projected or geographic.

M	*Projected by Universal Transverse Mercator system*
L	*Geographic (latitude and longitude)*

Easting, northing, zone

Record easting and northing UTM projected coordinates. Give a 6-figure easting, a 7-figure northing and a 2-figure grid zone (49–56 in Australia), to as high a degree of accuracy as the measurement method permits. In specific types of work in which high accuracy locations are needed (and suitable equipment is used), it may be appropriate to record sub-metre accuracy of locations, but this is not normal practice. In such cases, it is not recommended to be more precise than 0.1 m. Example:

Zone	Easting	Northing
55	692084	6094905

Latitude and longitude

Geographic coordinates may be recorded as degrees, minutes and seconds (DMS) or decimal degrees (DD). The latter provides greater utility for the data but five decimal places are required to obtain precision to the metre. Latitudes (giving the north/south part of the coordinate) will be negative in Australia. Example:

Latitude	Longitude
–35.27058	149.11181

Accuracy estimate (±)

Record the horizontal numerical accuracy of the coordinate location, preferably as a numeric value, but the categories below may also be used.

1	*<1 m*
2	*1–5 m*
3	*5–15 m*
4	*15–30 m*
5	*>30 m*

AIR PHOTO REFERENCE

The widespread availability of GNSS has reduced the need to identify location via air photo reference, but in some instances it may still be relevant. Refer to previous editions of the Field Handbook for detail regarding air photograph referencing.

LANDFORM

JG Speight, AJW Biggs and D Morand

LANDFORM DESCRIPTION

The description of landform has several purposes:

- It provides context to a point observation.
- The description is useful for finding relationships to support the extrapolation of point observations.
- It helps to determine suitable land uses and predict the land degradation that may follow various land uses.
- It has direct application to land use planning.

The landform description scheme that follows is intended to produce a record of observations rather than inferences. Where inference is implied in geomorphological terminology and practice, a clear record of what has been inferred is presented. While there is a focus on describing terrain using morphological characteristics, landscape processes and the genesis of landforms are implied in most cases. Similarly, the nature of the materials within a landform is generally secondary to the description, but in some cases there is a specific material type connotation – for example, karst landforms.

In this technique for describing landforms, the whole land surface is viewed as a mosaic of tiles of odd shapes and sizes. To impose order, the mosaic is treated as if the tiles are of two distinct sizes, the larger ones being comprised of mosaics of the smaller ones. The larger tiles, more than 600 m across, are called *landform patterns*. About 43 types of landform pattern are defined. They include, for example, flood plain, dunefield and hills. The smaller tiles, which form mosaics within landform patterns, are about 40 m or more across. These are called *landform elements*. More than 80 landform elements are defined – for example, dune, levee and playa.

While the descriptions of landform element and pattern have a local context, consideration should also be given to the broader view of landforms (as physiographic divisions, provinces and regions) in Australia, as described by Jennings and Mabbutt (1977) and more recently revised by Pain *et al.* (2011). There are 220 physiographic regions mapped in Australia.

Both landform elements and landform patterns may extend over areas larger than their characteristic dimensions and there is often a gradation from the smaller components

Table 1   Appropriate landform model for mapping at various scales

Map scale	Minimum width of mapping units	Appropriate landform model for mapping
1:500 000	1500 m	Landform pattern
1:250 000	750 m	Landform pattern
1:100 000	300 m	Landform pattern
1:50 000	150 m	Landform pattern
1:25 000	75 m	Landform pattern/landform element
1:10 000	30 m	Landform element
1:5 000	15 m	Landform element

(elements) into the larger (patterns) – for example, a plain within an alluvial plain. The shape of landform elements varies considerably and some are quite linear or curvilinear in nature. Their nominal diameter of 40 m should be viewed in the context of the nature of the element, such as a scarp, stream channel or levee.

Landform elements and landform patterns are described and classified into named types by the values of their *landform attributes*. Distinct suites of landform attributes relate to landform elements and landform patterns, respectively. Slope and position in a toposequence are key attributes for landform elements. Relief and stream occurrence are important in the determination of landform patterns.[2] The modes of geomorphological activity and geomorphological agent are relevant at both the element and pattern scale, thus genetic concepts are integral in the definition of many landforms.

Landform elements and patterns are an integral part of landscape units defined in the companion handbook *Guidelines for Surveying Soil and Land Resources* (McKenzie *et al.* 2008). A landform element is the landform part of a *land facet*, and a landform pattern is the landform part of a *land system*.

The depiction of landforms in maps can occur through a variety of methods. In the case of conventional polygonal maps, it is possible to depict either landform elements or landform patterns. Irrespective of scale, cartographic limits control the minimum size of a map unit (larger than about 3 mm on the map). As landform patterns have a characteristic dimension of about 600 m, this is the recommended size for sampling the landform pattern to evaluate its attributes. It is also the normal minimum width of a mapped landform pattern. It follows that landform patterns are best shown on a map at ~1:200 000 scale. Landform elements, with a characteristic dimension of about 40 m, are best shown at ~1:15 000 scale. Table 1 shows which of these two units is more appropriate on maps of various scales.

The advent of digital elevation models (DEMs) has led to the ability to depict landform features through a variety of methods, both in two and three dimensions. While it is possible to do so, quantitative analysis of digital elevation models to yield landform element and pattern *sensu stricto* is not yet common. Furthermore, digital elevation models are not a field tool. However, they are very valuable in determining the extent of either landform elements or patterns, particularly when the landform is obscured by vegetation.

---

2   Landform patterns and landform elements are formally defined in the abstract of a paper by Speight (1974) and are discussed in two other papers (Speight 1976, 1977).

Since many relationships between landforms and other phenomena occur at the landform element level, this model should be used to describe landform even when the scale of work dictates that only landform patterns can be *mapped*. In the field, describe both landform element and landform pattern.

An even smaller sampling area of 10 m in radius is convenient for field observation of certain attributes of landform and other features that are covered in the chapter 'Land surface'; see also 'The site concept'.

How much detail?

The attributes listed below are those required to distinguish between the types of landform given in the glossaries. The distinctions that have routinely been made in the past are likely to form a sound basis for survey practice.

For tasks where landform is of little concern, a very brief form of description is specified (pages 25 and 45). Some of the attributes are expressed in grade scales, with classes of even sizes, usually on a logarithmic base. Where more rigorous analysis is feasible, numerical values of attributes should be observed. Various additional attributes capable of precise quantification may be devised.

LANDFORM GENESIS

The two following sections on geomorphological modes and agents refer to the inferred *genesis* of a landform element or pattern. This genesis may have spanned thousands of years. Changes, such as erosion and aggradation, that are produced by current land use are separately assessed as attributes of the land surface.

To think clearly about the origin of a landform, one should ask two questions: 'What agent formed it?' and 'What was the mode of activity of that agent?'

Landforms created by different agents, such as wind, creep and stream flow, may result from the same mode of geomorphological activity – erosion, for example. Those created by the same agent may differ according to whether the mode of activity builds them up or erodes them down.

Mode of geomorphological activity

Various modes of geomorphological activity may be distinguished, as shown in Figure 1.

Gradational activity:

ER   *Eroded*

EA   *Eroded or aggraded*

AG   *Aggraded*

Anti-gradational activity:

HU   *Heaved up or elevated*

BU   *Built up*

EX   *Excavated or dug out*

SU   *Subsided or depressed*

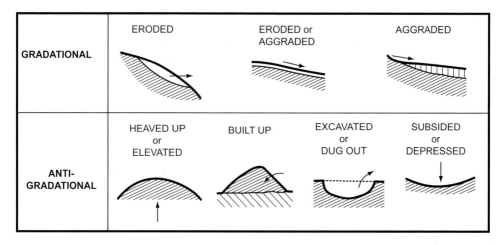

Figure 1   Modes of geomorphological activity.

Gradational activities are those that tend to reduce the land to a common elevation by removing material from higher places and depositing it in lower places (Chamberlin and Salisbury 1904, page 2), without necessarily reducing the angle of slope at every point. The work of streams and landslides is almost entirely in the gradational modes. However, this tendency is opposed by many processes that commonly act in an anti-gradational mode. These modes are characteristic of volcanism, diastrophism and various kinds of human and biological activity.

Many engineering works involve erosion and aggradation because these gradational modes use less energy than anti-gradational modes. For the same reason, erosion and aggradation may easily be induced unintentionally by land use.

To judge the mode of geomorphological activity responsible for a given landform, the observer must visualise a former surface that has suffered distortion, burial or removal of material, and seek evidence that such activity has taken place. Information on soil and substrate materials relevant to this investigation should be recorded as specified in other sections.

Allow for the recording of more than one mode of activity, together with options concerning geomorphological agents.

Geomorphological agent

Geomorphological agents that play a role in producing distinctive landforms are listed in Table 2. Much of the standard terminology relevant to landform elements presumes that the geomorphological agent responsible for a landform is known; this presumption has been explicitly incorporated in the following landform element glossary. In practice, the observer may find it difficult to infer correctly the agent responsible for producing a given landform element. The problem may be compounded by the apparent significance of more than one agent. In such cases, the observer should record both (a) the dominant agent, or the agent that is confidently inferred, and (b) a subordinate agent, or an agent that is dubiously inferred. In dubious cases, leave category (a) blank.

The importance of identifying landform elements with the agent 'channelled stream flow' is discussed under 'Channel depth relative to width' (page 39).

Table 2 Geomorphological agents significant for definition of landform elements and patterns

Agent		Description
Gravity	GR	Collapse, or particle fall
Precipitation	SO	Solution
	SM	Soil moisture status changes; creep
	WM	Water-aided mass movements; landslides
	SH	Sheet flow, sheet wash#, surface wash
Stream flow	OV	Over-bank stream flow, unchannelled floodplain flow
	CH	Channelled stream flow
Standing water	WA	Waves
	TI	Tides
	EU	Eustasy; changes in sea level
Groundwater	GR	Groundwater discharge/flow
Ice	FR	Frost, including freeze–thaw
	GL	Glacier flow
Wind	WI	Saltation
Internal forces	DI	Diastrophism; earth movements
	VO	Volcanism
Biological agents	BI	Non-human biological agents, e.g. coral, plants (in the case of peat)
	HU	Human agents
Extraterrestrial agents	IM	Impact by meteors

Hogg (1982) provides a description of the terms sheetflood, sheetwash and sheetflow, and proposed that the term sheetwash should be replaced by rainwash.

Underlying materials

While inferences about geomorphological agents and their mode of activity are essential to define many types of landform element (and landform pattern), observations of the underlying materials are not. Since these materials are often inaccessible to the observer, they should not be *definitive* for landforms. Landforms are seen as *indicators* of the underlying materials, permitting their extrapolation from limited exposures. The description of bedrock and regolith is discussed in the 'Substrate' chapter.

DESCRIPTION OF LANDFORM ELEMENT

A landform element may be described by the following attributes, assessed within a circle of about 20 m in radius (this dimension must be cognisant of the shape of the element):

- slope
- morphological type
- dimensions
- mode of geomorphological activity
- geomorphological agent.

These will establish most of the distinctions between landform elements that are implied by their geomorphological names. The glossary of types of landform element occurring in Australia refers explicitly to this set of attributes. A landform element that has been described may thus be assigned a type name. A shorter description consists simply of slope, morphological type and name (see page 25). Note that the dimension of a landform element may be larger than the site space.

Slope
Means of evaluation of slope

T   *Tripod-mounted instrument and staff*

A   *Abney level or clinometer and tape*

P   *Contour plan at 1:10 000 or larger scale*

E   *Estimate*

F   *Fine scale digital elevation model*

C   *Coarse scale digital elevation model*

Note that use of a digital elevation model is not a field measurement, but because DEMs may be used to derive a slope value for a site, codes for them have been included here. It is always preferable to measure slope in the field. The two scales of DEMs are defined as follows:

- Fine DEM: Slope derived from a DEM finer than 10 m resolution, calculated to give the effective slope of the line of steepest descent over an interval of 20 m centred on the observation (e.g. LiDAR, Air Photo-derived).
- Coarse DEM: Slope derived from a DEM with 10–30 m resolution, calculated to give the effective slope of the line of steepest descent over an interval of 20 m centred on the observation site (e.g. the national 1 second DEM and contour-based DEMs).

Slope value
Express slope tangent as a percentage using up to three significant figures – for example, 0.05%, 1%, 12.5%, 115%. Evaluate the slope over an interval of 20 m, straddling the point of soil observation (or centroid of the plot). Always observe and record the slope as precisely as the chosen survey method permits. The number of significant figures should relate to the accuracy of the method used to determine the slope. For example, an estimate is only accurate to integer values and a clinometer or abney level is only accurate to 0.5%. Slopes should only be reported to two decimal places if high precision equipment is used.

The observation should span no less than 20 m (page 5) so as not to be influenced too much by features of the microrelief that occur within the landform element.

Slope class
Slope classes are defined below as per Speight (1967, 1971). The optional word 'inclined' is used to distinguish slope from other attributes, for example 'gently inclined footslope' from 'gently undulating rises', and 'moderately inclined slope' from 'moderately spaced streams'.

The class boundaries given below and repeated in Table 4 (on page 37) are simply boundaries separating slope terms in common use, adjusted to regular logarithmic intervals. They do not refer to observed natural clustering of slope values, nor do they relate precisely to boundary criteria for land use, which vary arbitrarily between organisations and may change with advancing technology.

		Per cent	**Degrees**
LE	Level	<1%	<0.6
VG	Very gently inclined	1–3	0.6–1.7
GE	Gently inclined	3–10	1.7–6
MO	Moderately inclined	10–32	6–18
ST	Steep	32–56	18–29
VS	Very steep	56–100	30–45
PR	Precipitous	100–300	45–72
CL	Cliffed	>300	>72

The Level slope class can be further subdivided as required, as per FAO (2006).

		Per cent	**Degrees**
FL	Flat	0–0.2%	0–0.11
NF	Nearly flat	0.2–0.5%	0.11–0.29
NL	Nearly level	0.5–1%	0.29–0.57

It may sometimes be advantageous to split each of the classes 'very gently inclined', 'gently inclined' and 'moderately inclined' into two levels, the appropriate boundary values being 1.8%, 5.6% and 18%. There may also be compelling reasons for using other schemes of slope classes. However, schemes that do not have constant class widths from low to high slope values can lead to problems in subsequent statistical work.

Morphological type

Landform elements fall into 10 morphological types:

C	*Crest*
H	*Hillock*
R	*Ridge*
S	*Simple slope*
U	*Upper slope*
M	*Mid-slope*
L	*Lower slope*
F	*Flat*
V	*Open depression (vale)*
D	*Closed depression*

Of these, the types called 'slope' are also characterised by their inclination relative to adjacent elements as *waxing, waning, maximal* or *minimal*. Eight of the 10 types are shown in Figure 2.

Crests and *depressions* form the highest and lowest parts of the terrain. They are defined as follows:

Crest   Landform element that stands above all, or almost all, points in the adjacent terrain. It is characteristically smoothly convex upwards in downslope profile or in contour, or both. The margin of a crest element should be drawn at the limit of observed curvature.

Depression   Landform element that stands below all, or almost all, points in the adjacent terrain. A *closed depression* stands below all such points; an *open depression* extends at the same elevation, or lower, beyond the locality where it is observed. Many depressions are concave upwards and their margins should be drawn at the limit of observed curvature.

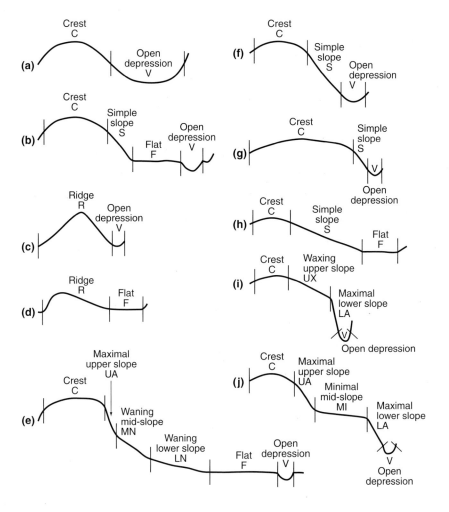

Figure 2 Examples of profiles across terrain divided into morphological types of landform element. Note that the boundary between crest and slope elements is at the end of the curvature of the crest. Each slope element is treated as if it were straight. Not shown are hillock or closed depression.

In any terrain, one may draw *slope lines* at right angles to the contour lines. Slope lines control the direction of many land-forming processes. In a terrain that has *relief*, each slope line runs from the extreme top (summit) of a *crest* down to the extreme bottom (lowest point) of a *closed depression* (Cayley 1859). Figure 3 shows many slope lines descending from several summits to one low point. The sequence of landform elements down a slope line is called a *toposequence*. Position in a toposequence is used to define morphological types of slope element that may occur between a crest and a depression. First the general type is defined:

Slope   Planar landform element that is neither a crest nor a depression and has an inclination greater than about 1%.

Landform elements that are slopes are treated as if each element is straight and meets another slope element at a slope break (Figure 2). Four morphological types are distinguished on their position in a toposequence relative to crests, flats (defined below) and depressions:

Simple slope   Slope element adjacent below a crest or flat and adjacent above a flat or depression.

Upper slope   Slope element adjacent below a crest or flat but not adjacent above a flat or depression.

Mid-slope   Slope element not adjacent below a crest or flat and not adjacent above a flat or depression.

Lower slope   Slope element not adjacent below a crest or flat but adjacent above a flat or depression.

A toposequence may include no slope element (Figures 2a, c, d), one simple slope (Figures 2b, f, g, h), or an upper slope and a lower slope (Figure 2i). All three cases occur in the area mapped in Figure 4. More complex toposequences may include an upper slope, a lower slope, and one or more mid-slopes (Figures 2e, j). The number of slope elements to distinguish depends either on the chosen level of survey detail or on observed differences in landform and their relationship to soil or vegetation.

Relative inclination of slope elements

Although lower slopes are often gentler than upper slopes, they need not be so (Figure 2i). A separate morphological attribute expresses the *relative inclination* of adjacent landform elements in a toposequence (crests and depressions are taken to be *gentler* than adjacent slopes).

X	*Waxing*	Element upslope is gentler, element downslope is steeper.
N	*Waning*	Element upslope is steeper, element downslope is gentler.
A	*Maximal*	Element upslope is gentler, element downslope is gentler.
I	*Minimal*	Element upslope is steeper, element downslope is steeper.

The morphological types *upper slope, mid-slope* and *lower slope* require two codes, as UX, MN, etc. (Figure 2), to include relative inclination. Other morphological types need no second code letter. Simple slopes are always maximal. For crests, flats, depressions, hillocks and ridges, relative inclination does not have a clear meaning.

Flats are defined as follows:

Flat   Planar landform element that is neither a crest nor a depression and is level or very gently inclined (less than 3% tangent approximately).

As defined, some flats and slopes may have the same inclination (1–3%). They differ in their typical relation to slope lines and toposequences. The slope line on a flat often runs parallel to the *course line* in a nearby open depression such as a stream channel. The slope line on a slope seldom does so, but makes an angle with the course line (Figure 3). A slope typically occurs in a toposequence from a crest to a depression. Where a flat occurs in such a toposequence (Figures 2b, e, h), it usually marks a change in process and a sharp change in the direction of the slope line. Most flats are in terrain with very little relief where crests do not occur.

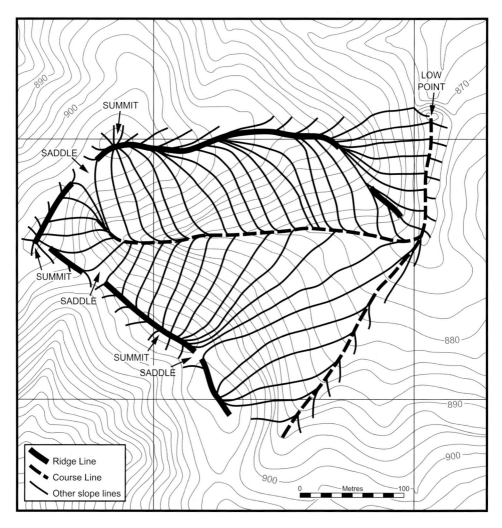

Figure 3   Slope lines overlaid on a contour map to show ridge lines and course lines where many slope lines come together.

Compound morphological types

Several types of landform feature have crests and adjoining slopes that are so small that a 20 m radius site would usually include both. Two compound morphological types are distinguished by the relative length of the crest:

Hillock   Compound landform element comprising a narrow crest and short adjoining slopes, the crest length being less than the width of the landform element.

Ridge   Compound landform element comprising a narrow crest and short adjoining slopes, the crest length being greater than the width of the landform element.

A *dune* is defined in the glossary as a hillock or ridge but, to allow for large dunes or detailed work, elements called *dunecrest* and *duneslope* are also defined. Other types of hillock and ridge may be divided into crest and slope elements if necessary.

Visualisation

When selecting a field site, visualise the set of morphological types of landform element that make up the landform pattern at that place. This includes placing boundaries between the elements. Then the site, or sites, can be properly located. Figure 4 shows an example of morphological types of landform element delineated in rolling low hills.

Dimensions

An occurrence of a landform element extends as far as its attributes remain constant. Its dimensions, which may be much greater than the specified sample area diameter of 40 m, can be significant to land use. Terms referring to dimension appear often in the definitions of landform element types. Three dimensions are distinguished, each to be expressed in metres:

Length   Horizontal distance between the upper and lower margins of the element, measured down a slope line. For crests, the slope line to be used is the ridge line; for depressions, the course line (see Figure 3). By this definition, many crests and open depressions become very long.

Width   Horizontal distance between the lateral margins of the element, measured perpendicularly to the length.

Height
(or depth)   Difference in elevation between the upper and lower margins of the element, measured along any slope line. Height can mean quite different things and must be carefully defined. For crests, ridges and hillocks, define the upper margin as the point where the selected slope line coalesces with others to form the ridge line. For depressions, define the lower margin as the point where the selected slope line coalesces with others to form the course line.

Location within the landform element

A site chosen to represent a landform element will often be placed centrally within it. For various reasons, a site may not be centrally placed and this should be recorded. The

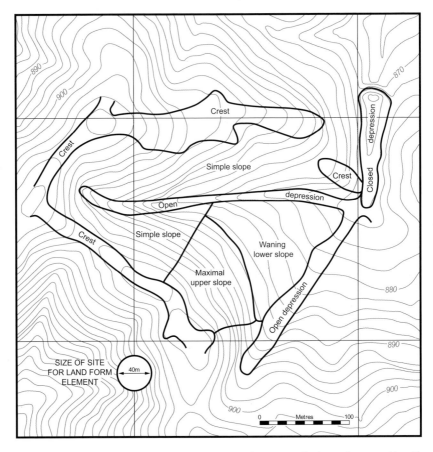

Figure 4 A landform pattern of rolling low hills mapped into morphological types of landform element. Note that the crests and depressions in this case are mainly narrower than the recommended site size.

vertical position of the site within the *height* of the landform element may be the best measure:

T   *Top third of the height of the landform element*
M   *Middle third of the height of the landform element*
B   *Bottom third of the height of the landform element*

Location within a toposequence

For detailed work, the location of the site within the toposequence down a slope line may relate to landform processes. Unfortunately, slope lines, by definition, extend from a summit to the lowest point (Cayley 1859), which may be many kilometres apart. One must arbitrarily determine the *effective* top and bottom of the toposequence. These effective end points come where the slope line coalesces with other slope lines to form a ridge line or a course line[3] (Figure 3). In practice, ridge lines and course lines are to be excluded from the toposequence (see entry for height in the earlier section on dimensions). One arbitrary rule that will exclude the ridge line is to put the top of the toposequence where the contour curvature exceeds 60° in 40 m. The course line may be excluded by the same rule, or by putting the bottom of the toposequence at a stream channel.

Any site can be located by its vertical and horizontal distances from the defined top and bottom of the toposequence. Drainage height is one of these measures.

The toposequence concept leads to definition of the attributes *specific catchment area* and *specific dispersal area* (Speight 1974, 1980) that predict hillslope hydrology and erosion (see, for example, Moore *et al.* 1988).

Short description of a landform element

Slope class, morphological type and a name of the landform element type from the glossary form the briefest description that is likely to be useful. Examples:

	Slope class	Morphological type and relative inclination	Element
Gentle crest: summit surface	GE	C	SUS
Gentle waxing upper slope: (no name)	GE	UX	
Precipitous maximal mid-slope: scarp	PR	MA	SCA
Steep waning lower slope: cliff-footslope	ST	LN	CFS
Gentle waning lower slope: footslope	GE	LN	FOO
Very steep maximal lower slope: (no name)	VS	LA	
Very gentle open depression: drainage depression	VG	V	DDE
Moderate hillock: tor	MO	H	TOR
Level ridge: levee	LE	R	LEV
Very gentle flat: valley flat	VG	F	VLF

In each case, the name of the landform element type implies that other, unstated attributes have been observed or inferred. These other attributes are given below. Their values are stated in the glossary of landform elements and in the key (Table 3).

LANDFORM ELEMENT KEY AND GLOSSARY

The key (Table 3) and glossary below aims to provide a concise set of names for types of landform element. Where different landform elements in a survey area have the same type name, distinguish them by qualifying terms based on attributes of landform element or of land surface. Examples are:

- steep maximal upper slope
- rocky, gentle upper slope
- severely gullied footslope

Each glossary definition is based on the attributes that have been listed. Attribute values that have not been observed but merely inferred from a glossary definition must not be treated as data.

---

3   To define ridge lines and course lines by slope line coalescence, as shown in Figure 3, departs from the concept of Cayley (1859). He defined them as those slope lines that intersect at a knot (that is, a saddle/pass).

Table 3   Key to landform element types

Morphological type		Mode of activity	Land-forming agent	Other discriminators	Landform element type	
Crest	C	Eroded	Creep or sheet wash (SM, SH)	Not very wide, steeper	Crest	HCR
				Very wide, gentler	Summit surface	SUS
				Below an adjoining ridge	Saddle	SAD
		Built up or eroded	Wind (WI)		Dunecrest	DUC
Hillock	H	Eroded	Creep or sheet wash (SM, SH)	With bare rock	Tor	TOR
				Regolith-covered	Residual rise	RER
		Heaved up	Volcanism (VO)		Tumulus	TUM
		Built up or eroded	Wind (WI)	(see also ridge)	Dune*	DUN
				Weakly oriented	Hummocky dune	DUH
				Crescentic	Barchan dune	DUB
				Parabolic	Parabolic dune	DUP
				Longitudinal	Linear or longitudinal (seif) dune	DUF
			Volcanism (VO)		Cone (volcanic)	CON
			Groundwater, biological (GR, BI)		Spring mound	SPM
			People (HU)		Mound	MOU
Ridge	R	Built up or eroded	Over-bank flow (OV)		Levee	LEV
			Channel flow (CH)		Bar (stream)	BAR
				Relict bar	Scroll	SCR
			Channel or over-bank flow (CH, OV)	Relict levee	Prior stream	PST
			Wind (WI)	(see also hillock)	Dune*	DUN
			Wind (WI)	From adjacent playa	Lunette	LUN
			Wind or waves (WI, WA)	From beach; relict	Beach ridge	BRI
				From adjacent beach	Foredune	FOR
			Glacier flow (GL)	Associated with a cirque	Arete	ARE
			People (HU)	(see also slope)	Embankment*	EMB
				To enclose a depression	Dam	DAM
Slope (unspecified: upper, mid-, lower, or simple)	S, U, M, L	Eroded	Collapse	Cliffed, very wide, maximal	Cliff	CLI

Morphological type	Mode of activity	Land-forming agent	Other discriminators	Landform element type	
		Collapse, landslide or sheet wash	Precipitous, very wide, maximal	Scarp	SCA
		Sheet wash, creep or landslide		Slope	HSL
	Eroded and aggraded	People (HU)		Cut face	CUT
		Landslide	Hummocky	Landslide	LDS
	Built up	People (HU)	(see also ridge)	Embankment*	EMB
Simple slope — S	Eroded or aggraded	Channel flow (CH)	Very wide	Bank (stream)	BAN
	Built up or eroded	Waves (WA)		Beach	BEA
		Wind (WI)		Duneslope	DUS
Mid-slope — M	Eroded	Collapse, landslide or sheet wash (GR, WM, SH)	Small scarp and scarp-footslope together	Breakaway	BRK
		Landslide or sheet wash (WM, SH)	At foot of a cliff (see also lower slope)	Cliff-footslope*	CFS
		Creep or sheet wash (SM, SH)	At foot of a scarp (see also lower slope)	Scarp-footslope*	SFS
	Eroded or aggraded	Any agent	Minimal slope	Bench	BEN
		People (HU)	Minimal slope	Berm (i) #	BER
Lower slope — L	Eroded	Landslide or sheet wash (WM, SH)	At foot of a cliff (see also mid-slope)	Cliff-footslope*	CFS
		Creep or sheet wash (SM, SH)	At foot of a scarp (see also mid-slope)	Scarp-footslope*	SFS
	Eroded or aggraded	Sheet wash (SH)	Large, gentle, mainly eroded (see also flat)	Pediment*	PED
		Sheet wash, landslide or creep (SH, WM, SM)	Waning slope, not large	Footslope	FOO
			Rock fragments	Talus	TAL
	Aggraded	Collapse (GR)	Mainly formed by erosion; aggradation is local	Cliff-footslope*	CFS
Flat — F	Any mode	Any agent	Large, gentle, mainly eroded (see also flat)	Plain	PLA
	Eroded	Sheet wash (SH)	Rock	Rock flat	RFL
		Waves (WA)	Rock	Rock platform	RPL

Table 3 (cont.)

Morphological type	Mode of activity	Land-forming agent	Other discriminators	Landform element type	
		People (HU)		Cut-over surface	COS
		Glacier flow (GL)	Bare, grooved	Glacial pavement	GLP
	Eroded or dug out	Wind or sheet wash (WI, SH)	Soil-eroded, small	Scald	SCD
	Eroded or aggraded	Sheet wash (SH)	Large, gentle, unidirectional, mainly eroded (see also lower slope)	Pediment*	PED
	Eroded or aggraded	Channel flow (CH)	Radial, mainly aggraded	Fan	FAN
		Channel or over-bank flow (CH, OV)	Enclosed by slopes, mainly aggraded	Valley flat	VLF
			Relict, small	Terrace flat	TEF
			At channel margin, small	Channel bench	CBE
	Aggraded	Over-bank flow (OV)	Large	Backplain	BKP
		Channel flow (CH)	Large	Scroll plain	SRP
		Channel or over-bank flow (CH, OV)	Radial, on a flood plain	Flood-out	FLD
			Relict, large	Terrace plain	TEP
		Tides (TI)	Undifferentiated	Tidal flat	TDF
			Permanently inundated	Subtidal flat	SBF
			Frequently inundated	Intertidal flat	ITF
			Infrequent inundation	Supratidal flat	STF
			Rarely inundated	Extratidal flat	ETF
	Built up	People (HU)		Fill-top	FIL
		Waves (WA)	Above a beach	Berm (ii) #	BER
		Coral (BI)		Reef flat	REF
Open depression V	Eroded	Landslide, creep or sheet wash (WM, SM, SH)	Sloping, short	Alcove	ALC
		Channel flow and collapse (CH, GR)	With precipitous walls	Gully	GUL
		Glacier flow (GL)	Part dug out and closed depression	Cirque*	CIR
	Eroded or aggraded	Sheet wash (SH)	Gentle or flat, long	Drainage depression	DDE
	Eroded, aggraded, dug out or built up	Channel flow (CH)		Stream channel	STC

Morphological type	Mode of activity	Land-forming agent	Other discriminators	Landform element type	
			Secondary to the main stream channel	Anabranch	ANA
			Mainly eroded; part of stream channel	Stream bed	STB
		Channel flow and tides (CH, TI)	Tapered; tide water only	Tidal creek	TDC
	Aggraded	Over-bank flow (etc.) (OV)	Tapered; river and tide water	Estuary	EST
			Flat; surface watertable (see also closed depression)	Swamp*	SWP
		Channel flow (CH)	Between scrolls	Swale (ii) #	SWL
		Wind or waves (WI, WA)	Between ridges	Swale (i) #	SWL
	Built up or dug out	People (HU)		Trench	TRE
Closed depression	Any mode	Any agent	Large, water-filled	Lake	LAK
D			Large, usually dried up	Playa	PLY
			Small, water filled	Swamp	SWP
	Eroded	Wind (WI)		Deflation basin	DBA
		Solution (SO)		Solution doline	DOL
		Collapse (GR)		Collapse doline	DOC
		Channel flow (CH)	Long, curved	Ox-bow	OXB
		Waves or coral (WA, BI)	Large, saltwater-filled	Lagoon	LAG
	Aggraded	Over-bank flow or peat (OV, BI)	Surface watertable (see also open depression)	Swamp*	SWP
	Dug out	Wind (WI)	Small	Blow-out	BOU
		Glacier flow (GL)	Partly eroded open depression	Cirque*	CIR
		Volcanism (VO)	Usually water-filled	Maar	MAA
		Volcanism, meteor or people (VO, IM, HU)	By explosion	Crater	CRA
		People (HU)		Pit	PIT

*Landform element type name occurs more than once.
See the element definition.

29

Glossary

ALC	*Alcove*	Moderately inclined to very steep, short open depression with concave cross-section, eroded by collapse, landslides, creep or surface wash.
ANA	*Anabranch*	A section of *stream channel* that diverts from the main channel and rejoins at a point further downstream.
ARE	*Arete*	A ridge created by glacial action, separating two *cirques*.
BAN	*Bank (stream bank)*	Very short, very wide slope, moderately inclined to precipitous, forming the marginal upper parts of a *stream channel* and resulting from erosion or aggradation by channelled stream flow.
BAR	*Bar (stream bar)*	Elongated, gently to moderately inclined, low ridge built up by channelled stream flow; part of a *stream bed*. See also *scroll*.
BEA	*Beach*	Short, low, very wide slope, gently or moderately inclined, built up or eroded by waves, forming the shore of a *lake* or sea.
BEN	*Bench*	Short, gently or very gently inclined minimal mid-slope element eroded or aggraded by any agent.
BER	*Berm*	(i) Short, very gently inclined to level minimal mid-slope in an *embankment* or *cut face*, eroded or aggraded by human activity. (ii) Flat built up by waves above a *beach*.
BKP	*Backplain*	Large flat resulting from aggradation by over-bank stream flow at some distance from the *stream channel* and in some cases biological (peat) accumulation; often characterised by a high watertable and the presence of *swamps* or *lakes*; part of a covered plain landform pattern.
	Billabong	See *ox-bow*.
BOU	*Blow-out*	Usually small, open or closed depression excavated by the wind, typically in a dune.
BRI	*Beach ridge*	Very long, nearly straight, low ridge, built up by waves and usually modified by wind. A beach ridge is often a relict feature remote from the *beach*. Typically comprised of a sand-sized fraction (siliceous sand, shell).
BRK	*Breakaway*	Steep maximal mid-slope or upper slope, generally comprising both a very short *scarp* (free face) that is often bare rockland, and a stony *scarp-footslope* (debris slope); often standing above a *pediment*.
	Channel	See *stream channel*.
CBE	*Channel bench*	Flat at the margin of a *stream channel* aggraded and in part eroded by over-bank and channelled stream flow; an incipient flood plain. Channel benches have been referred to as 'low terraces', but the term terrace should be restricted to landform patterns above the influence of active stream flow.
CFS	*Cliff-footslope*	Slope situated below a *cliff*, with its contours generally parallel to the line of the *cliff*, eroded by sheet wash or water-aided mass movement, and aggraded locally by collapsed material from above.

CIR	*Cirque*	Precipitous to gently inclined, typically closed depression of concave contour and profile excavated by ice. The closed part of the depression may be shallow, the larger part being an open depression like an *alcove*.
	Claypan	See *deflation basin*.
CLI	*Cliff*	Very wide, cliffed (greater than 72°) maximal slope usually eroded by gravitational fall as a result of erosion of the base by various agencies; sometimes built up by marine organisms (cf. *scarp*).
CON	*Cone (volcanic)*	Hillock with a circular symmetry built up by volcanism. The crest may form a ring around a *crater*.
COS	*Cut-over surface*	Flat excavated by human activity.
CRA	*Crater*	Steep to precipitous closed depression excavated by explosions due to volcanism, human action or impact of an extraterrestrial object.
CUT	*Cut face*	Slope excavated by human activity.
DAM	*Dam*	Ridge built up by human activity so as to close a depression to hold water.
DBA	*Deflation basin*	Basin excavated by wind erosion which removes loose material, commonly above a resistant or wet layer (sometimes referred to as claypans, not to be confused with a *playa*).
DDE	*Drainage depression*	Level to gently inclined, long, narrow, shallow open depression with smoothly concave cross-section, rising to moderately inclined side slopes, eroded or aggraded by sheet wash.
DOC	*Collapse doline*	Steep-sided, circular or elliptical closed depression, commonly funnel-shaped, characterised by subsurface drainage and formed by collapse of underlying caves within bedrock.
DOL	*Solution doline*	Steep-sided, circular or elliptical closed depression, commonly funnel-shaped, characterised by subsurface drainage and formed by dissolution of the surface or underlying bedrock.
DUB	*Barchan dune*	Crescent-shaped *dune* with tips extending leeward (downwind), making this side concave and the windward (upwind) side convex. Barchan dunes tend to be arranged in chains extending in the dominant wind direction. Typically comprised of a sand-sized fraction (siliceous sand).
DUC	*Dunecrest*	Crest built up or eroded by the wind (see *dune*), when the dune is large enough to be a pattern.
DUF	*Linear or longitudinal (seif) dune*	Sharp-crested, elongated, longitudinal (linear) *dune* or chain of *dunes*, oriented parallel, rather than transverse (perpendicular), to the prevailing wind. Typically comprised of a sand-sized fraction (siliceous sand, shell). Not to be confused with the trailing arms of parabolic dunes.
DUH	*Hummocky (weakly oriented) dune*	Very gently to moderately inclined rises or hillocks built up or eroded by wind and lacking distinct orientation or regular pattern. Typically comprised of a sand-sized fraction (siliceous sand, shell).

DUN	*Dune*	Moderately inclined to very steep ridge or hillock built up by the wind. This element may comprise *dunecrest* and *duneslope*. Typically comprised of a sand-sized fraction (siliceous sand, shell).
DUP	*Parabolic dune*	*Dune* with a long, scoop-shaped form, convex in the downwind direction so that its horns point upwind, whose ground plan approximates the form of a parabola. The dunes left behind can be referred to as trailing arms. Where many such dunes have traversed an area, these can give the appearance of linear dunes. Typically comprised of a sand-sized fraction (siliceous sand, shell).
DUS	*Duneslope*	Slope built up or eroded by the wind (see *dune*).
EMB	*Embankment*	Ridge or slope built up by human activity.
EST	*Estuary*	*Stream channel* close to its junction with a sea or *lake*, where the action of channelled stream flow is modified by tide and waves. The width typically increases downstream.
ETF	*Extratidal flat*	Flat that is rarely subject to inundation by saline tidal water other than exceptional storm or cyclonic tides.
FAN	*Fan*	Large, gently inclined to level element with radial slope lines inclined away from a point, resulting from aggradation (or occasionally from erosion) by channelled, often braided, stream flow, or possibly by sheet flow.
FIL	*Fill-top*	Flat aggraded by human activity.
FLD	*Flood-out*	Flat inclined radially away from a point on the margin or at the end of a *stream channel*, aggraded by over-bank stream flow, or by channelled stream flow associated with channels developed within the over-bank flow; part of a covered plain landform pattern.
FOO	*Footslope*	Moderately to very gently inclined waning lower slope resulting from aggradation or erosion by sheet flow, earth flow or creep (cf. *pediment*).
FOR	*Foredune*	Very long, nearly straight, moderately inclined to very steep ridge built up by the wind from material from an adjacent *beach*. Typically comprised of a sand-sized fraction (siliceous sand, shell).
GLP	*Glacial pavement*	A typically bare rock surface that has been scraped by the action of glaciers.
GUL	*Gully*	Open depression with short, precipitous walls and moderately inclined to very gently inclined floor, eroded by channelled stream flow and consequent collapse and water-aided mass movement.
HCR	*Crest*	Very gently inclined to steep crest, smoothly convex, eroded mainly by creep and sheet wash. A typical element of mountains, hills, low hills and rises.
HSL	*Slope*	Gently inclined to precipitous slope, commonly simple and maximal, eroded by sheet wash, creep or water-aided mass movement. A typical element of mountains, hills, low hills and rises, and less commonly on plains.

ITF	*Intertidal flat*	Flat subject to regular, saline tidal inundation of mostly high frequency.
LAG	*Lagoon*	Closed depression filled with water that is typically saline or brackish, bounded at least in part by forms aggraded or built up by waves or reef-building organisms.
LAK	*Lake*	Large water-filled closed depression.
LDS	*Landslide*	Moderately inclined to very steep slope, eroded in the upper part and aggraded in the lower part by water-aided mass movement, characterised by irregular hummocks.
LEV	*Levee*	Very long, low, narrow, nearly level, sinuous ridge immediately adjacent to a *stream channel*, built up by over-bank flow. Levees are built, usually in pairs bounding the two sides of a *stream channel*, at the level reached by frequent floods. For an artificial levee, use *embankment*. For a relict levee, see *scroll*. See also *prior stream*.
LUN	*Lunette*	Elongated, gently curved, low ridge built up by wind on the margin of a *playa*, typically with a moderate, wave-modified slope towards the *playa* and a gentle outer slope.
MAA	*Maar*	Level-floored, commonly water-filled closed depression with a nearly circular, steep rim, excavated by volcanism interacting with groundwater.
MOU	*Mound*	Hillock built up by human activity.
OXB	*Ox-bow*	Long, curved, commonly water-filled closed depression eroded by channelled stream flow, but closed as a result of aggradation by channelled or over-bank stream flow during the formation of a meander plain landform pattern. The floor of an ox-bow may be more or less aggraded by over-bank stream flow, wind and biological (peat) accumulation.
PED	*Pediment*	Very gently inclined to moderately inclined waning lower slope, with slope lines inclined in a single direction, or somewhat convergent or divergent, eroded or sometimes slightly aggraded by sheet flow (cf. *footslope*). It is underlain by bedrock.
PIT	*Pit*	Closed depression excavated by human activity.
PLA	*Plain*	Large, very gently inclined or level element, of unspecified geomorphological agent or mode of activity.
PLY	*Playa*	Shallow, level-floored closed depression, intermittently water-filled, but mainly dry due to evaporation, bounded as a rule by flats aggraded by sheet flow and channelled stream flow.
PST	*Prior stream*	Long, generally sinuous, low ridge built up from materials originally deposited by stream flow along the line of a former *stream channel*. The landform element may include a depression marking the old stream bed, and relict *levees*.
REF	*Reef flat*	Flat built up to sea level by marine organisms, typically comprised of coral and/or rubble.

RER	*Residual rise*	Hillock of very low to extremely low relief (<30 m) and very gentle to steep slopes. This term is used to refer to an isolated rise surrounded by other landforms such as plains.
RFL	*Rock flat*	Flat of bare consolidated rock, usually eroded by sheet wash.
RPL	*Rock platform*	Flat of consolidated rock eroded by waves.
SAD	*Saddle*	A low point (dip) along a ridge or the lowest point between two adjacent elevated areas. In one axis along the ridge or between the elevated areas the land slopes up in both directions and in the other axis, typically at right angles to the first, the land slopes down in both directions.
SBF	*Subtidal flat*	Flat below mean low water springs.
SCA	*Scarp*	Very wide, steep to precipitous maximal slope eroded by gravity, water-aided mass movement or sheet flow (cf. *cliff*).
SCD	*Scald*	Flat, bare of vegetation, from which soil has been eroded by surface wash or wind.
SCR	*Scroll*	Long, curved, very low ridge built up by channelled stream flow and left relict by channel migration. Part of a meander plain landform pattern.
SFS	*Scarp-footslope*	Waning or minimal slope situated below a *scarp*, with its contours generally parallel to the line of the *scarp*.
SPM	*Spring mound*	Small to large mound, often steep-sided, formed by the discharge of groundwater, often with associated biological aggradation. May include very small closed depressions.
SRP	*Scroll plain*	Large relict flat resulting from aggradation by channelled stream flow as a stream migrates from side to side; the dominant element of a meander plain landform pattern. This landform element may include occurrences of *scroll*, *swale* and *ox-bow*.
STB	*Stream bed*	Linear, generally sinuous open depression forming the bottom of a *stream channel*, eroded and locally excavated, aggraded or built up by channelled stream flow. Parts that are built up include *bars*.
STC	*Stream channel*	Linear, generally sinuous open depression, in parts eroded, excavated, built up and aggraded by channelled stream flow. This element comprises *stream bed* and *banks*.
STF	*Supratidal flat*	Flat subject to infrequent inundation by saline tidal water.
SUS	*Summit surface*	Very wide, level to gently inclined crest with abrupt margins, commonly eroded by water-aided mass movement or sheet wash.
SWL	*Swale*	(i) Linear, level-floored open depression excavated by wind, or left relict between ridges built up by wind or waves (e.g. dunes), or built up to a lesser height than them. (ii) Long, curved, open or closed depression left relict between *scrolls* built up by channelled stream flow.

SWP	*Swamp*	Almost level, closed or almost closed depression with a seasonal or permanent watertable at or above the surface, commonly aggraded by over-bank stream flow and sometimes biological (peat) accumulation (cf. *lake*).
TAL	*Talus*	Moderately inclined or steep waning lower slope, consisting of rock fragments (scree) aggraded by gravity.
TDC	*Tidal creek*	Intermittently water-filled open depression in parts eroded, excavated, built up and aggraded by channelled tide-water flow; type of *stream channel* characterised by a rapid increase in width downstream.
TDF	*Tidal flat*	Flat subject to inundation by water that is usually saline or brackish, aggraded by tides. See also *intertidal flat* (ITF), *supratidal flat* (STF), *extratidal flat* (ETF) and *subtidal flat* (SBF).
TEF	*Terrace flat*	Small flat aggraded or eroded by channelled or over-bank stream flow, standing above a *scarp* and no longer frequently inundated; a former *valley flat* or part of a former flood plain.
TEP	*Terrace plain*	Large or very large flat aggraded by channelled or over-bank stream flow, standing above a *scarp* and no longer frequently inundated; part of a former flood plain.
TOR	*Tor*	Steep to precipitous hillock, typically convex, with a surface mainly of bare rock, either coherent or comprising subangular to rounded large boulders (exhumed core-stones, also themselves called tors) separated by open fissures; eroded by sheet wash or water-aided mass movement.
TRE	*Trench*	Open depression, narrower than long, excavated by human activity.
TUM	*Tumulus*	Hillock heaved up by volcanism (or, elsewhere, built up by human activity at a burial site).
VLF	*Valley flat*	Small, gently inclined to level flat, aggraded or sometimes eroded by channelled or over-bank stream flow, typically enclosed by *slopes*; a miniature alluvial plain landform pattern.

DESCRIPTION OF LANDFORM PATTERN

The significant kinds of landform pattern in Australia may be described and differentiated by the following attributes assessed within a circle of about 300 m in radius:

- relief
- modal slope
- stream channel occurrence
- mode of geomorphological activity
- geomorphological agent
- status of geomorphological activity
- component landform elements.

The glossary of landform pattern types that follows is based explicitly on these attributes. Many other attributes may be observed, particularly by means of remote sensing (Speight 1977), so permitting finer discrimination between landform patterns. In fact, landform pattern description is seldom built up from field observations alone, so that this section is marginal to the purpose of this Field Handbook. It aims, rather, to provide a secure broader geomorphological context for field work.

In the field, the observer should take care not to include parts of adjacent dissimilar landform patterns and thereby compromise the description of the landform pattern in which the observation point is found. Landform pattern boundaries, such as slope–floodplain junctions or dissection heads, may be recorded by a diagram.

Relief

Relief is defined as the difference in elevation between the high and low points of a land surface. Its estimation will be made easier by visualising two surfaces of accordance that are planar or gently curved, one touching the major crests of a landform pattern, and the other passing through the major depressions. The average vertical separation of the two surfaces is a measure of the relief. Make this estimation at a field site, either visually or by using a map, and express it in metres.

Relief is the definitive characteristic for the terms *mountains, hills, low hills, rises* and *plains* when used as types of erosional landform pattern (Table 4). The class boundaries, shown in Table 4 and Table 5, are set at 300 m, 90 m, 30 m and 9 m. These class limits and the class names are similar to those used by Löffler (1974), and are broadly compatible with those of Löffler and Ruxton (1969).

Modal slope

Modal slope is defined as the most common class of slope occurring in a landform pattern. Where slope classes have been obtained by systematic sampling, define the classes using equal increments on a scale of the logarithm of the slope tangent, a procedure intended to normalise frequency distributions of observed slope (Speight 1971). Where the most common slope class is estimated by direct observation, it is thought that the estimate will compare with that calculated using the log-normal model.

Modal slope class determines the use of certain adjectives applied to landform patterns that are characterised by alternating crests and depressions. These are: *rolling* for moderate modal slopes (10–32%); *undulating* for gentle slopes (3–10%); and *gently undulating* for very gentle slopes (1–3%) (compare with Soil Survey Staff 1951, pages 161–165). The other slope classes – *precipitous, very steep, steep* and *level* – are to be applied as they stand. The terminology for simple erosional landform patterns based on relief and modal slope is given in Table 4. The table defines the category *badlands* by various combinations of high slope values and low relief values. These combinations imply extremely close spacing of streams or valleys. Specifically, if one assumes a sawtooth terrain profile, the valley spacing implied is less than 100 m in areas with 50 m relief and less than 30 m in areas with 5 m relief; these values appear to accord with usage.

Table 5 lists types of landform pattern defined in the glossary according to their typical relief class.

Table 6 lists types of landform pattern in order of their typical class of modal slope. This table should not be regarded as definitive, for slope within each type of landform pattern may vary widely.

Table 4 Simple types of erosional landform pattern characterised by relief and modal slope

	Modal terrain slope						
	LE Level <1% (about 1:300)	**VG** Very gently inclined 1–3% (about 2%)	**GE** Gently inclined 3–10% (about 6%)	**MO** Moderately inclined 10–32% (about 20%)	**ST** Steep 32–56% (about 40%)	**VS** Very steep 56–100% (about 70%)	**PR** Precipitous >100% (about 150%)
Relief							
M Very high >300 m (about 500 m)	—	—	—	**RM** Rolling mountains	**SM** Steep mountains	**VM** Very steep mountains	**PM** Precipitous mountains
H High 90–300 m (about 150 m)	—	—	**UH** Undulating hills	**RH** Rolling hills	**SH** Steep hills	**VH** Very steep hills	**PH** Precipitous hills
L Low 30–90 m (about 50 m)	—	—	**UL** Undulating low hills	**RL** Rolling low hills	**SL** Steep low hills	**VL** Very steep low hills	**B** Badlands
R Very low 9–30 m (about 15 m)	—	**GR** Gently undulating rises	**UR** Undulating rises	**RR** Rolling rises	**SR** Steep rises	**B** Badlands	**B** Badlands
P Extremely low <9 m (about 5 m)	**LP** Level plain	**GP** Gently undulating plain	**UP** Undulating plain	**RP** Rolling plain	**B** Badlands	**B** Badlands	**B** Badlands

Table 5 Landform pattern types ordered by typical relief class

Those types for which the relief class is definitive are in italics

Typical relief	Landform patterns
Very high >300 m	*Mountain*, volcano
High 90–300 m	*Hill*, volcano, caldera, meteor crater
Low 30–90 m	*Low hill*, volcano, caldera, meteor crater
Very low 9–30 m	*Rise*, terrace, dunefield, lava plain, coral reef, peneplain, karst
Extremely low <9 m	*Plain*, pediment, pediplain, sheet-flood fan, alluvial fan, alluvial plain, flood plain, meander plain, bar plain, covered plain, anastomotic plain, stagnant alluvial plain, delta, playa plain, tidal flat, beach ridge plain, chenier plain, sand plain, made land

Table 6 Landform pattern types ordered by typical modal slope class

Those types for which the relief class is definitive are in italics

Typical modal slope class	Landform pattern types
Precipitous >100%	(Rare in Australia)
Very steep 56–100%	*Mountain*, escarpment, volcano, caldera
Steep 32–56%	*Hill*
Moderately inclined 10–32%	*Low hill*, karst, meteor crater
Gently inclined 3–10%	*Rise*, beach ridge plain, dunefield, coral reef
Very gently inclined 1–3%	*Pediment*, alluvial fan, sand plain, lava plain
Level <1%	*Plain*, sheet-flood fan, pediplain, peneplain, alluvial plain, flood plain, meander plain, bar plain, covered plain, anastomotic plain, stagnant alluvial plain, terrace, tidal flat, made land, playa plain

Stream channel occurrence

A number of attributes describing the occurrence and pattern of surface stream channels have diagnostic value. Use of the following attributes may clarify the observable differences between landform patterns, particularly in plains where mapping criteria are elusive.

When assessing attributes of stream channel occurrence, it is easy to make errors by not setting limits to the area to be described. Tentative landform pattern boundaries must be drawn to clarify these limits. Major stream channels are best mapped as wholly within one landform pattern or another, rather than marking a boundary.

Stream channel spacing

The average spacing of stream channels, L/N, is determined by counting the number, N, of their intersections with an arbitrary line of length L.[4]

A convenient tool for estimating channel spacing is a circle, with a circumference of 2 km at map or photo scale. Suitable classes for stream channel spacing, based on existing data, are:

AB *Absent or very rare* >2500 m

SP *Sparse* 1585–2500 m

---

4 The average spacing, L/N, is the reciprocal of *stream channel frequency*, N/L (Speight 1977), a measure advocated by McCoy (1971) to replace the less convenient *drainage density, Dd* (Horton 1945). Mark (1974) has demonstrated a logical and empirical relationship from which it follows that stream channel spacing is related to drainage density by: $L/N = 1.571/Dd$

VW	*Very widely spaced*	1000–1585 m
WS	*Widely spaced*	625–1000 m
MS	*Moderately spaced*	400–625 m
CS	*Closely spaced*	250–400 m
VC	*Very closely spaced*	158–250 m
NU	*Numerous*	<158 m

Stream channel development

The degree of development of stream channels may be categorised as follows:

O	*Absent*	No traces of channelled flow can be detected.
I	*Incipient*	Traces of channelled flow are very shallow, narrow and discontinuous.
E	*Erosional*	Continuous linear channels occur; their width and depth are considerable and display somewhat constant values suited to the available flow. Flood plains are not formed.
A	*Alluvial*	Continuous linear channels occur, with rather large width and depth; they are essentially constant with downstream distance and are suited to the available flow. Flood plains of vertical or lateral accretion are formed.

Channel depth relative to width

Channel depth and width refer to the dimensions of a landform that is dominated by channelled stream flow. The limit of channelled stream flow dominance must be identified before width or depth can be estimated. Depth is taken from the top of the stream bank down to the average height of the line following the deepest part of the channel.

The distinction between stream bank and slope or scarp according to dominant process requires particular care where streams are incised, especially if they are cut into terraces that could be mistaken for flood plains. For detailed studies, keep records of width and depth measurements. In other surveys, use the following classes of relative depth.

D	*Deep*	Width/depth ratio <20:1
M	*Moderately deep*	Width/depth ratio 20:1 to 50:1
S	*Shallow*	Width/depth ratio 50:1 to 150:1
V	*Very shallow*	Width/depth ratio >150:1

Stream channel migration

The presence of relict channel landforms or unvegetated, newly formed channel margins or immovable channel margins may permit an assessment of channel migration as:

R	*Rapidly migrating*
S	*Slowly migrating*
F	*Fixed*

Stream-wise channel pattern

In a traverse downstream, it may happen that tributaries enter the stream at frequent intervals, or that the stream splits into distributaries, or that these tendencies are absent (the *non-tributary* case) or are in balance with each other (the braided or anastomotic case called here *reticulated*) giving four classes of stream-wise channel pattern:

T	*Tributary*
N	*Non-tributary*
D	*Distributary*
R	*Reticulated*

(See Figure 5a.)

Channel network integration

In an *integrated* channel network, one can traverse from any point on a stream channel to any other point on a stream channel without passing through any landform elements other than stream channels. The channel network may be *interrupted* at points where water loss into the ground or the atmosphere is sufficiently large, and in the extreme case, typical of karst terrain, the surface stream network is *disintegrated*. Classes of channel network integration are:

I	*Integrated*
P	*Interrupted (partial integration)*
D	*Disintegrated*

(See Figure 5b.)

Channel network directionality

This attribute combines two simpler attributes: the degree of lineation, that is, the degree to which the channels tend to align in an organised way; and the degree of convergence or divergence of channels in the downstream direction. The latter is distinct from tributary/distributary behaviour, which refers to the combining and splitting of stream channels, rather than their directionality. Classes are:

F	*Centrifugal*	maximum divergence >90°
D	*Divergent*	maximum divergence between 10° and 90°
U	*Unidirectional*	convergence or divergence <10°
C	*Convergent*	maximum convergence between 10° and 90°
P	*Centripetal*	maximum convergence >90°
B	*Bidirectional*	two lineations (for example 'trellis')
N	*Non-directional*	no significant orientation, convergence or divergence

(See Figure 5c.)

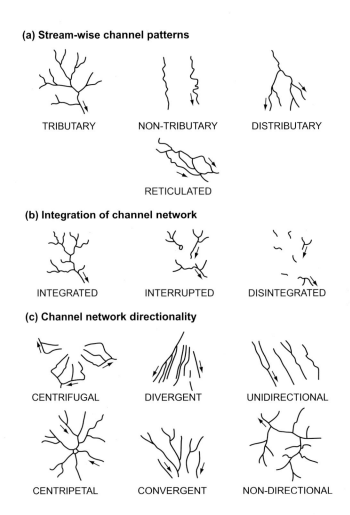

Figure 5 Stream channel pattern attributes.

To illustrate the significance of stream channel occurrence in discriminating between landform patterns, Table 7 presents examples of landform pattern types ordered according to each of these seven attributes.

Mode of geomorphological activity
The modes of geomorphological activity are those considered in the description of landform elements (see page 15). Table 8 indicates the dominant mode of geomorphological activity in common types of landform pattern.

Geomorphological agent
Landform patterns are subject to the same geomorphological agents as are landform elements (see page 16). The problems of assigning agents and of expressing the relative significance of

Table 7 Examples of types of landform pattern ordered according to attributes of stream channel occurrence

Attributes of stream channel occurrence	Examples of landform pattern types
Stream channel spacing	
Absent or very rare	Sand plain, beach ridge plain
Sparse	Made land
Very widely spaced	Very steep mountains
Widely spaced	Meander plain, steep hills
Moderately spaced	Anastomotic plain, undulating rises
Closely spaced	Steep low hills
Very closely spaced	Precipitous hills
Numerous	Badlands, bar plain, pediment
Stream channel development	
Absent	Dunefield, pediplain, playa plain
Incipient	Pediment, sheet-flood fan
Erosional	Mountains, hills, rises
Alluvial	Meander plain, bar plain, covered plain
Stream channel depth relative to width	Covered plain, anastomotic plain
Deep	Meander plain
Moderately deep	Bar plain
Shallow	Pediment
Very shallow	
Stream channel migration	
Rapidly migrating	Bar plain, meander plain
Slowly migrating	Covered plain
Fixed	Mountains, hills, rises
Stream-wise channel pattern	
Tributary	Mountains, hills, rises
Non-tributary	Meander plain, covered plain
Distributary	Delta, sheet-flood fan, pediment
Reticulated	Bar plain, anastomotic plain
Stream channel network integration	Mountains, hills, rises
Integrated	Volcano
Interrupted (partial integration)	Karst
Disintegrated	
Stream channel network directionality	Volcano
Centrifugal	Pediment, sheet-flood fan
Divergent	Meander plain, bar plain, covered plain
Unidirectional	Hills, rises
Convergent	Caldera
Centripetal	Mountains, hills, rises
Non-directional	Hills, rises
Bi-directional	

more than one agent for a landform pattern are even more acute than for landform elements. Make provision for listing dominant, co-dominant and accessory geomorphological agents.

Table 9 shows the incidence of geomorphological agents in types of landform pattern. Landform patterns, being larger than landform elements, commonly have longer histories. The

Table 8 Landform pattern types grouped according to the dominant mode of geomorphological activity

Dominant mode of geomorphological activity		Landform pattern types
Gradational		
ER	*Eroded*	Mountains, hills, rises, karst, pediplain, peneplain
EA	*Eroded or aggraded*	Pediment, made land
AG	*Aggraded*	Alluvial plain, flood plain, alluvial fan, bar plain, meander plain, covered plain, terrace, sheet-flood fan, lava plain, playa plain, tidal flat
Anti-gradational		
HU	*Heaved up*	Marine plain
BU	*Built up*	Volcano, coral reef, dunefield, beach ridge plain
EX	*Excavated*	Caldera, meteor crater
SU	*Subsided*	(Rare in Australia)

Table 9 Incidence of geomorphological agents in types of landform pattern

		Landform patterns		
	Geomorphological agent	**Dominant agent**	**Co-dominant agent**	**Accessory agent**
GR	***Gravity*** Collapse, or particle fall	(Rare in Australia)	Stock	Hills, karst, volcano, dunefield, meteor crater
	Precipitation			
SO	Solution	Karst		Badlands
SM	Soil moisture status changes; creep		Hills	Playa plain
WM	Water-aided mass movements; landslides		Hills, stock	Badlands
SH	Sheet flow, sheet wash, surface wash	Hills, sheet-flood fan, pediment, pediplain, peneplain	Playa plain	Karst, badlands
	Stream flow			
OV	Over-bank stream flow, unchannelled	Covered plain, anastomotic plain	Flood plain, alluvial plain, terrace	
CH	Channelled stream flow	Meander plain, bar plain	Flood plain, alluvial plain, terrace	Delta
WI	Wind	Dunefield	Playa plain, beach ridge plain, sand plain	Beach ridge plain, pediment
	Groundwater			Karst
	Ice			
FR	Frost, including freeze–thaw	(Rare in Australia)		Hills
GL	Glacier flow	Glacial valley, moraine		

Table 9 (cont.)

		Landform patterns		
	Geomorphological agent	**Dominant agent**	**Co-dominant agent**	**Accessory agent**
	Standing water			
WA	Waves	Lacustrine plain	Beach ridge plain, playa plain, marine plain	Tidal flat, delta
TI	Tides	Tidal flat, island, coral island, coral reef	Marine plain	Beach ridge plain, delta
EU	Eustasy: changes in sea level	(Rare in Australia)		
	Internal forces			
DI	Diastrophism: earth movements	(Rare in Australia)		
VO	Volcanism	Volcano, caldera, lava plain	Erosion caldera	
	Biological agents			
BI	Non-human biological agents; coral	Coral reef		
HU	Human agents	Made land		
	Extraterrestrial agents			
IM	Impact by meteors	Meteor crater		

landform pattern description often identifies longer acting or relict geomorphological modes and agents.

Status of geomorphological activity

It is important for theoretical and practical purposes to distinguish, if possible, between landform patterns in which the formative geomorphological processes continue at the present time, and those in which they are no longer active, the landform features being relict. The problem in assigning activity status is that many processes are episodic, so that the observation of no activity may mean that an episodic process is in a quiescent phase. The following scale does not distinguish between processes that operate continuously but extremely slowly and those episodic processes that are very rare:

C   *Continuously active*

F   *Frequently active*

S   *Seldom active*

B   *Barely active to inactive*

R   *Relict*

U   *Unspecified*

Table 10 shows how types of landform pattern vary in their status of geomorphological activity. Note that flood plains, including bar plains, meander plains, covered plains, anastomotic plains and deltas, are distinguished from terraces or stagnant alluvial plains by

Table 10 Typical activity status of the dominant geomorphological agent in types of landform pattern

Typical activity status	Landform patterns
Continuously active	Mountains, hills, rises, karst, coral reef
Frequently active	Pediment, sheet-flood fan, flood plain, bar plain, meander plain, covered plain, anastomotic plain, alluvial fan, tidal flat, dunefield, playa plain
Seldom active	(lower) Terrace
Barely active to inactive	Pediplain, peneplain, stagnant alluvial plain
Relict	Caldera,* volcano, shield volcano, meteor crater, (higher) terrace, beach ridge plain, chenier plain, lava plain, made land
Unspecified	Plain, alluvial plain

* In Australia.

having frequently active rather than seldom active or inactive stream flow. This may have legal significance. The frequency of occurrence of inundation that is classed as *frequently active* in this Field Handbook is an Average Recurrence Interval of 50 years or less.

A landform pattern may change from one type to another type if the status of geomorphological activity changes for any reason, including human interference such as diverting a stream or building a dam.

Component landform elements

Certain kinds of landform element are typical of a given type of landform pattern. Others are found commonly and others occasionally in a given type. These landform elements are listed for each type of landform pattern in the glossary.

Short description of a landform pattern

The categories of relief and modal slope class given by the code letters in the margins of Table 4, added to a name from the glossary, form the briefest description of landform patterns that is likely to be useful (examples below). Named landform pattern types are discriminated by many other attributes, some of which are given below. Type names must be used with great care. In the glossary, cross-references are given to the similar landform pattern types with which a given type could be confused.

Description	Relief	Modal slope	Pattern
Extremely low, very gently inclined sand plain	P	VG	SAN
Low, gently inclined lava plain	L	GE	LAV
High, steep (relict) volcano	H	ST	VOL

LANDFORM PATTERN GLOSSARY

The definitions in this glossary refer explicitly to the attributes of landform patterns that have been set down in the preceding sections. Consequently, they differ from the original definitions by the authors who are cited.

Cross-references and the tables in this section should be used to distinguish between landform pattern types that are similar. The relief-related landform patterns (plain, rise, low

Table 11   Classification of riverine landform patterns

Low or very low relief	
More than one plain level	Terraced land (alluvial)
One plain level (seldom active or relict)	Terrace (alluvial)
Extremely low relief	
Undifferentiated	Alluvial plain
Inactive or barely active	Stagnant alluvial plain
Frequently active	
in sea or lake	Delta
elsewhere	
undifferentiated	Flood plain
differentiated	Bar plain
	Meander plain
	Covered plain
	Anastomotic plain

Table 12   Discrimination between flood plains

	Type of flood plain			
Attributes	**Bar plain**	**Meander plain**	**Covered plain**	**Anastomotic plain**
Stream channel				
Spacing	Numerous	Widely spaced	Widely spaced	Moderately spaced
Depth/width	Shallow	Moderately deep	Deep	Deep
Migration	Rapid	Rapid	Slow	Slow
Stream-wise pattern	Reticulated	Non-tributary	Non-tributary	Reticulated
Network directionality	Unidirectional	Unidirectional	Unidirectional	Divergent/ unidirectional
Mode of geomorphological activity				
	Eroded/ aggraded	Eroded/ aggraded	Aggraded	Aggraded
Geomorphological agent				
Dominant	Channelled stream flow	Channelled stream flow	Over-bank stream flow	Over-bank stream flow
Minor		Over-bank stream flow		Channelled stream flow
Landform elements				
Scroll plain		Dominant		
Backplain			Dominant	Dominant
Stream channel	Typical	Typical	Typical	Typical
stream bed	Typical	Typical	Typical	Typical
bar	Dominant	Typical		
bank		Typical		
Scroll		Typical		
Levee			Typical	Typical
Swamp			Common	Common
Ox-bow		Common		

hill, hill, mountain) should be used as the starting point in describing landform in the field. It will often be necessary to review other data (topography, geology) in order to ascertain the correct landform pattern.

Riverine landform patterns comprise a hierarchical classification as shown in Table 11. Four types of flood plain differ in many ways, as set out in Table 12. Alluvial fan, sheet-flood fan and pediment are particularly difficult to distinguish. They differ mainly in that stream channels are better developed and deeper on alluvial fans, and that pediments are almost entirely erosional while fans are depositional.

Glossary

ALF   *Alluvial fan*   Level to gently inclined, complex landform pattern of extremely low relief. The rapidly migrating stream channels are shallow to moderately deep, locally numerous, but elsewhere widely spaced. The channels form a centrifugal to divergent, integrated, reticulated to distributary pattern. The landform pattern includes areas that are *bar plains*, being aggraded or eroded by frequently active channelled stream flow, and other areas comprising *terraces* or *stagnant alluvial plains* with slopes that are greater than usual, formed by channelled stream flow but now relict. Incision in the upslope area may give rise to an erosional stream bed between scarps.
Typical elements: stream bed, bar, plain.
Common element: scarp.
Compare with *sheet-flood fan* and *pediment*.

ALP   *Alluvial plain*   Level landform pattern with extremely low relief. The shallow to deep alluvial stream channels are sparse to widely spaced, forming a unidirectional, integrated network. There may be frequently active erosion and aggradation by channelled and over-bank stream flow, or the landforms may be relict from these processes.
Typical elements: stream channel (stream bed and bank), plain (dominant).
Common elements: bar, scroll, levee, backplain, swamp.
Occasional elements: ox-bow, flood-out, lake.
Included types of landform pattern are: *flood plain, bar plain, meander plain, covered plain, anastomotic plain, delta, stagnant alluvial plain, terrace, terraced land*.

ANA   *Anastomotic plain*   *Flood plain* with slowly migrating, deep alluvial channels, usually moderately spaced, forming a divergent to unidirectional, integrated reticulated network. There is frequently active aggradation by over-bank and channelled stream flow.
Typical elements: stream channel (stream bed and bank), levee, backplain (dominant).
Common element: swamp.
Compare with other types under *alluvial plain* and *flood plain*.

BAD   *Badlands*   Landform pattern of low to extremely low relief (less than 90 m) and steep to precipitous slopes, typically with numerous fixed, erosional stream channels which form a non-directional, integrated tributary network. There is continuously active erosion by collapse, landslide, sheet flow, creep and channelled stream flow.
Typical elements: ridge (dominant), stream bed or gully.
Occasional elements: summit surface, crest, slope, talus.
Compare with *mountains, hills, low hills, rises* and *plain*.

BAR   *Bar plain*   *Flood plain* with numerous rapidly migrating, shallow alluvial channels forming a unidirectional, integrated reticulated network. There is frequently active aggradation and erosion by channelled stream flow (described by Melton 1936).
Typical elements: stream bed, bar (dominant).
Compare with other types under *alluvial plain* and *flood plain*.

BEA   *Beach ridge plain*   Level to gently undulating landform pattern of extremely low relief on which stream channels are absent or very rare; it consists of relict, parallel beach ridges.
Typical elements: beach ridge (co-dominant), swale (co-dominant).
Common elements: beach, foredune, tidal creek.
Compare with *chenier plain*.

CAL   *Caldera*   Rare landform pattern typically of very high relief and steep to precipitous slope. It is without stream channels or has fixed, erosional channels forming a centripetal, integrated tributary pattern. The landform was excavated by volcanism (evacuation of a magma chamber) (cf. *volcano*).
Typical elements: scarp, slope, lake.
Occasional elements: cone, crest, stream channel.

CHE   *Chenier plain*   Level to gently undulating landform pattern of extremely low relief on which stream channels are very rare. The pattern consists of relict, parallel, linear ridges built up by waves, separated by, and built over, flats (mud flats) aggraded by tides or over-bank stream flow.
Typical elements: beach ridge (co-dominant), flat (co-dominant).
Common elements: tidal flat, swamp, beach, foredune, tidal creek.
Compare with *beach ridge plain*.

COI   *Coral island*   Landform pattern of extremely low relief, typically without stream channels and surrounded by water. The landform is built up and modified by wave and wind action and biological activity.
Typical elements: beach, berm, lagoon, reef flat.
Common elements: beach ridge, foredune, swale.
Occasional element: swamp.

COR   *Coral reef*   Continuously active or relict landform pattern built up to the sea level of the present day or of a former time by corals and other organisms. It is mainly level, with moderately inclined to precipitous slopes below the sea level. Stream channels are generally absent, but there may occasionally be fixed, deep, erosional tidal stream channels forming a disintegrated non-tributary pattern.
Typical elements: reef flat, lagoon, cliff (submarine).
Common elements: beach, beach ridge.

COV   *Covered plain*   *Flood plain* with slowly migrating, deep alluvial channels, usually widely spaced and forming a unidirectional, integrated non-tributary network. There is frequently active aggradation by over-bank stream flow (described by Melton 1936).
Typical elements: stream channel (stream bed and bank), levee, backplain (dominant).
Common element: swamp.
Compare with other types under *alluvial plain* and *flood plain*.

DEL   *Delta*   *Flood plain* projecting into a sea or lake, with slowly migrating, deep alluvial channels, usually moderately spaced, typically forming a divergent, integrated distributary network. This landform is aggraded by frequently active over-bank and channelled stream flow that is modified by tides.
Typical elements: stream channel (stream bed and bank), levee, backplain (co-dominant), swamp (co-dominant), lagoon (co-dominant).
Common elements: beach ridge, swale, beach, estuary, tidal creek.
Compare with other types under *alluvial plain, flood plain* and *chenier plain*.

DUN   *Dunefield*   Level to rolling landform pattern of very low or extremely low relief without stream channels, built up or locally excavated, eroded or aggraded by wind.
Typical elements: dune or dunecrest, duneslope, swale, blow-out, crest, residual rise, slope.
Common elements: hummocky dune, barchan dune, parabolic dune, linear dune.
Included types of landform pattern are: *longitudinal dunefield, parabolic dunefield*.

ERC   *Erosion caldera*   Rare landform pattern typically of high relief and steep to precipitous slope. It is without stream channels or has fixed, erosional channels. The landform is a caldera or relict volcano that has been eroded both inwards and outwards, often resulting in loss of part of the crest and sides.
Typical elements: scarp, slope.
Occasional elements: crest, stream channel.

ESC *Escarpment* Steep to precipitous landform pattern forming a linearly extensive, straight or sinuous, inclined surface, which separates terrains at different altitudes; a *plateau* is commonly above the escarpment. Relief within the landform pattern may be high (hilly) or low (planar). The upper margin is often marked by an included cliff or scarp.
Typical elements: crest, slope, cliff-footslope.
Common elements: cliff, scarp, scarp-footslope, talus, footslope, alcove.
Occasional element: stream bed.

FLO *Flood plain* *Alluvial plain* characterised by frequently active erosion and aggradation by channelled or over-bank stream flow. Unless otherwise specified, 'frequently active' is to mean that flow has an Average Recurrence Interval of 50 years or less.
Typical elements: stream channel (stream bed and bank), plain (dominant).
Common elements: bar, scroll, levee, backplain, swamp.
Occasional elements: ox-bow, flood-out, scroll.
Included types of landform pattern are: *bar plain, meander plain, covered plain, anastomotic plain.*
Related relict landform patterns are: *stagnant alluvial plain, terrace, terraced land* (partly relict).

GLV *Glacial valley* Valley created wholly or in part by the action of current or historical glaciers. Typically straight and U-shaped in cross-section. Stream channels unidirectional.
Typical elements: slope, drainage depression, residual rise.
Common element: stream channel.
Occasional elements: glacial pavement, cirque, arete.

HIL *Hills* Landform pattern of high relief (90–300 m) with gently inclined to precipitous slopes. Fixed, shallow, erosional stream channels, closely to very widely spaced, form a non-directional or convergent, integrated tributary network. There is continuously active erosion by wash and creep and, in some cases, rarely active erosion by landslides.
Typical elements: crest, slope (dominant), drainage depression, stream bed.
Common elements: footslope, alcove, valley flat, gully.
Occasional elements: tor, summit surface, scarp, landslide, talus, bench, terrace, doline.
Compare with *mountains, low hills, rises* and *plain.*

ISL *Island* Landform pattern of extremely low to high relief, typically without stream channels, and surrounded by water. The landform is built up by volcanism and modified by wind, waves, sheet flow and stream flow.
Typical elements: beach, slope, cliff, rock platform.
Common elements: beach ridge, foredune, footslope, crest, swamp, tor.
Occasional elements: lagoon, tidal creek, tidal flat.

KAR	*Karst*	Landform pattern of unspecified relief and slope (for specification use the terms in Table 4, such as 'Karst rolling hills'), typically with fixed, deep, erosional stream channels forming a non-directional, disintegrated tributary pattern and many closed depressions without stream channels. It is eroded by continuously active solution and rarely active collapse, the products being removed through underground channels.

Typical elements: crest, slope (dominant), doline.
Common elements: summit surface, valley flat, plain, alcove, drainage depression, stream channel, scarp, footslope, landslide.
Occasional element: talus.

LAC	*Lacustrine plain*	Level landform pattern with extremely low relief formerly occupied by a lake but now partly or completely dry. It is relict after aggradation by waves and by deposition of material from suspension and solution in standing water. The pattern is usually bounded by wave-formed features such as cliffs, rock platforms, beaches, berms and lunettes. These may be included or excluded.

Typical element: plain.
Common elements: beach, cliff.
Occasional elements: rock platform, berm.
Compare with *playa plain*.

LAV	*Lava plain*	Level to undulating landform pattern of very low to extremely low relief typically with widely spaced, fixed, erosional stream channels that form a non-directional, integrated or interrupted tributary pattern. The landform pattern is aggraded by volcanism (lava flow) that is generally relict; it is subject to erosion by continuously active sheet flow, creep and channelled stream flow.

Typical elements: plain, slope, stream bed, bench, drainage depression.
Occasional elements: tumulus, cone, residual rise, swamp.

LON	*Longitudinal dunefield*	*Dunefield* characterised by long, narrow sand dunes and wide, flat swales. The dunes are oriented parallel with the direction of the prevailing wind, and in cross-section one slope is typically steeper than the other.

Typical elements: dune or dunecrest, duneslope, swale, blow-out.
Compare with *parabolic dunefield*.

LOW	*Low hills*	Landform pattern of low relief (30–90 m) and gentle to very steep slopes, typically with fixed, erosional stream channels, closely to very widely spaced, which form a non-directional or convergent, integrated tributary pattern. There is continuously active sheet flow, creep and channelled stream flow.

Typical elements: crest, slope (dominant), drainage depression, stream bed.
Common elements: footslope, alcove, valley flat, gully.
Occasional elements: tor, summit surface, landslide, doline.
Compare with *mountains, hills, rises* and *plain*.

MAD *Made land* Landform pattern typically of very low or extremely low relief and with slopes either level or very steep. Sparse, fixed, deep, artificial stream channels form a non-directional, interrupted tributary pattern. The landform pattern is eroded and aggraded, and locally built up or excavated, by human agency.

Typical elements: fill-top (dominant), cut-over surface, cut face, embankment, berm, trench.

Common elements: mound, pit, dam.

MAR *Marine plain* *Plain* eroded or aggraded by waves, tides or submarine currents, and aggraded by deposition of material from suspension and solution in sea water, elevated above sea level by earth movements or eustasy, and little modified by subaerial agents such as stream flow or wind.

Typical element: plain.

Occasional elements: dune, stream channel.

MEA *Meander plain* *Flood plain* with widely spaced, rapidly migrating, moderately deep alluvial stream channels which form a unidirectional, integrated non-tributary network. There is frequently active aggradation and erosion by channelled stream flow with subordinate aggradation by over-bank stream flow (described by Melton 1936).

Typical elements: stream channel (stream bed, bank and bar), scroll, scroll plain (dominant).

Common element: ox-bow.

Compare with other types under *alluvial plain* and *flood plain*.

MET *Meteor crater* Rare landform pattern comprising a circular closed depression (see crater landform element) with a raised margin; it is typically of low to high relief and has a large range of slope values, without stream channels, or with a peripheral integrated pattern of centrifugal tributary streams. The pattern is excavated, heaved up and built up by a meteor impact and is now relict.

Typical elements: crater (scarp, talus, footslope and plain), crest, slope.

MOR *Moraine*[5] Uncommon landform pattern comprising gentle slopes of low relief with variable stream networks. Channelled stream flow, incision, erosion and aggradation may be common. The pattern is created by aggradation of material (till) mobilised by glaciers. Sheet flow, creep and channelled stream flow may be common.

Typical element: slope (dominant).

Common element: stream channel.

Associated elements: cirque, arete.

---

5   Only extant as relict features in Australia

MOU	*Mountains*	Landform pattern of very high relief (greater than 300 m) with moderate to precipitous slopes and fixed, erosional stream channels that are closely to very widely spaced and form a non-directional or diverging, integrated tributary network. There is continuously active erosion by collapse, landslide, sheet flow, creep and channelled stream flow. Typical elements: crest, slope (dominant), stream bed. Common elements: talus, landslide, alcove, valley flat, scarp. Occasional elements: cirque, footslope. Compare with *hills, low hills, rises* and *plain*.
PAR	*Parabolic dunefield*	*Dunefield* characterised by sand dunes with a long, scoop-shaped form, convex in the downwind direction so that its trailing arms point upwind; the ground plan, when developed, approximates the form of a parabola. Where many parabolic dunes have been active, the trailing arms give the impression of a longitudinal dunefield. Typical elements: dune or dunecrest, duneslope, swale, blow-out. Compare with *longitudinal dunefield*.
PED	*Pediment*	Very gently to moderately inclined landform pattern of extremely low relief, typically with numerous rapidly migrating, very shallow incipient stream channels, which form a centrifugal to diverging, integrated reticulated pattern. It is underlain by bedrock, eroded, and locally aggraded, by frequently active channelled stream flow or sheet flow, with subordinate wind erosion. Pediments characteristically lie downslope from adjacent hills with markedly steeper slopes. Includes covered, mantled and rock pediments. See Twidale (1981, 2014). Typical elements: pediment, plain, stream bed. Compare with *sheet-flood fan* and *alluvial fan*.
PEP	*Pediplain*	Level to very gently inclined landform pattern with extremely low relief and no stream channels, eroded by barely active sheet flow and wind. Largely relict from more effective erosion by stream flow in incipient stream channels as on a *pediment*. Formed from the coalescence of multiple pediments in semi-arid to arid environments (described by King 1953). Typical element: plain.
PLA	*Plain*	Level to undulating or, rarely, rolling landform pattern of extremely low relief (less than 9 m). Compare with *mountains, hills, low hills* and *rises*.
PLT	*Plateau*	Level to rolling landform pattern of *plains*, *rises* or *low hills* standing above a cliff, scarp or *escarpment* that extends around a large part of its perimeter. A bounding scarp or cliff landform element may be included or excluded; a bounding *escarpment* would be an adjacent landform pattern. Typical elements: plain, summit surface, cliff. Common elements: crest, slope, drainage depression, rock flat, scarp. Occasional element: stream channel.

PLY	*Playa plain*	Level landform pattern with extremely low relief, typically without stream channels, aggraded by rarely active sheet flow and modified by wind, waves and soil phenomena. Typical elements: playa, lunette, plain. Compare with *lacustrine plain*.
PNP	*Peneplain*	Level to gently undulating landform pattern with extremely low relief and sparse, slowly migrating alluvial stream channels which form a non-directional, integrated tributary pattern. It is eroded by barely active sheet flow, creep, and channelled and over-bank stream flow (described by Davis 1889). Formed by the down-wearing of a large surface. Typical elements: plain (dominant), stream channel.
RII	*River island (eyot)*	Landform pattern of extremely low to low relief, typically without stream channels, surrounded by water (in a river). The landform is created by streamflow erosion of either consolidated or unconsolidated land and is modified by wind, waves, sheet flow and stream flow. Typical elements: tidal flat(s), slope, crest. Common elements: beach, footslope, tidal creek, swamp. Occasional elements: scarp, prior stream, levee.
RIS	*Rises*	Landform pattern of very low relief (9–30 m) and very gentle to steep slopes. The fixed, erosional stream channels are closely to very widely spaced and form a non-directional to convergent, integrated or interrupted tributary pattern. The pattern is eroded by continuously active to barely active creep and sheet flow. Typical elements: crest, slope (dominant), footslope, drainage depression. Common elements: valley flat, stream channel. Occasional elements: gully, fan, tor. Compare with *mountains, hills, low hills* and *plain*.
SAI	*Sand island*	Landform pattern of extremely low to low relief, typically without stream channels and surrounded by water. The landform is built up and modified by wave action and biological activity. Typical elements: beach, berm, foredune, swale, dune. Common elements: beach ridge, swamp. Occasional elements: lagoon, tidal creek, tidal flat.
SAN	*Sand plain*	Level to gently undulating landform pattern of extremely low relief and without channels; formed possibly by sheet flow or stream flow, but now relict and modified by wind action. Typical element: plain. Occasional elements: dune, playa, lunette.
SHF	*Sheet-flood fan*	Level to very gently inclined landform pattern of extremely low relief with numerous rapidly migrating, very shallow incipient stream channels forming a divergent to unidirectional, integrated or interrupted reticulated pattern. The pattern is aggraded by frequently active sheet flow and channelled stream flow, with subordinate wind erosion. Typical elements: plain, stream bed. Compare with *alluvial fan* and *pediment*.

SHV	*Shield volcano (relict)*	Volcanic landform of low relief and gently inclined slopes with centrifugal stream channels.
		Typical element: slope (dominant).
		Common elements: crest, pediment, footslope.
		Occasional elements: gully, stream channel, maar, bench.
STA	*Stagnant alluvial plain*	*Alluvial plain* on which erosion and aggradation by channelled and over-bank stream flow is barely active or inactive because of reduced water supply, without apparent incision or channel enlargement that would lower the level of stream action.
		Typical elements: stream channel (stream bed and bank), plain (dominant).
		Common elements: bar, scroll, levee, backplain, swamp.
		Occasional elements: ox-bow, flood-out, lake.
		Compare with *flood plain* and *terrace*.
STO	*Stock*	A remnant of the vent of a volcano or plutonic body with an areal extent less than 100 km^2 (USGS). Typically high to very high and very steep to cliffed landform pattern without stream channels, or with erosional stream channels forming a centrifugal, interrupted tributary pattern. The landform is built up by volcanism and is modified by erosional agents.
		Typical elements: cliff, tor, talus.
		Occasional elements: summit surface, gully.
TEL	*Terraced land (alluvial)*	Landform pattern including one or more *terraces* and often a *flood plain*. Relief is low or very low (9–90 m). Terrace plains or terrace flats occur at stated heights above the top of the stream bank.
		Typical elements: terrace plains, terrace flats, scarps, scroll plain, stream channel.
		Occasional elements: stream channel, scroll, levee.
TER	*Terrace (alluvial)*	Former *flood plain* on which erosion and aggradation by channelled and over-bank stream flow is barely active or inactive because deepening or enlargement of the stream channel has lowered the level of flooding. A pattern that has both a former *flood plain* and a significant, active *flood plain*, or that has former *flood plains* at more than one level, becomes *terraced land*.
		Typical elements: terrace plain (dominant), scarp, channel bench.
		Occasional elements: stream channel, scroll, levee.
TID	*Tidal flat*	Level landform pattern with extremely low relief and slowly migrating, deep alluvial stream channels, which form non-directional, integrated tributary patterns; it is aggraded by frequently active tides.
		Typical elements: plain (dominant), intertidal flat, supratidal flat, stream channel.
		Occasional elements: lagoon, dune, beach ridge, beach.

VOL   *Volcano*   Typically very high and very steep landform pattern without stream
            (relict)   channels, or with erosional stream channels forming a centrifugal,
                          interrupted tributary pattern. The landform is built up by volcanism
                          and is modified by erosional agents.
                          Typical elements: cone, crater.
                          Common elements: scarp, crest, slope, stream bed, lake, maar.
                          Occasional element: tumulus.

VEGETATION

D Lewis, E Addicott, N Cuff, A Kitchener, D Lynch,
K Zdunic, A West, J Balmer and B Sparrow

The scope of this chapter is to outline the field standards required for collecting vegetation attributes commonly measured and recorded at field sites to describe and classify Australian vegetation types. A *vegetation type* is defined here as a community that has a floristically uniform structure and composition, often described by its dominant species (Meagher 1991). A *vegetation unit*, by comparison, is a spatial category which contains a vegetation type or group of co-occurring vegetation types (ESCAVI 2003). The attributes contained in this chapter include both core attributes and those commonly recorded in the field. Core attributes are considered essential to describe and classify a vegetation type, including:

- strata (layering of vegetation – defined by growth form and height characteristics)
- structure (height, cover)
- growth form (physiognomy)
- floristics (species composition, and structure).

The structural characteristics, including growth form, are collected at both the site scale (i.e. within a plot), and when appropriate at the species level (per taxon). For example, within a plot, what is the percentage canopy cover of the upper stratum? For a species recorded in the upper stratum, what percentage canopy cover does the species contribute within the plot (across all strata/substrata)?

The Field Handbook acknowledges that Australian state and territory agencies have existing survey protocols in place. Therefore, depending on the location field sampling is conducted, also refer to the most recent field survey protocols, as provided in Table 13. The field standards outlined here should be consistent across the continent; however, the methods employed may vary from state to state. This is to ensure field data collected using standardised methods are comparable across jurisdictional boundaries and easily linked for national or other research purposes.

Table 13 Vegetation field survey manuals for Australian states and territories

National – TERN	White *et al.* (2012) Sparrow *et al.* (2020) TERN (2023)	https://www.tern.org.au/wp-content/uploads/TERN-Rangelands-Survey-Protocols-Manual_web.pdf https://doi.org/10.3389/fevo.2020.00157 https://www.tern.org.au/emsa-protocols-manual
Australian Capital Territory	Baines *et al.* (2013) ACT Government (2015)	https://www.environment.act.gov.au/__data/assets/pdf_file/0009/576846/CPR_Technical_Report_28.pdf https://www.environment.act.gov.au/__data/assets/pdf_file/0005/728600/Schedule-1-Environmental-offsets-Assessment-Methodology-FINAL-2.pdf
New South Wales	Sivertsen (2009) BioNet	https://www.environment.nsw.gov.au/resources/nativeveg/10060nvinttypestand.pdf https://www.environment.nsw.gov.au/topics/animals-and-plants/biodiversity/nsw-bionet/about-bionet-vegetation-classification
Northern Territory	Brocklehurst *et al.* (2007)	https://hdl.handle.net/10070/635994 https://depws.nt.gov.au/rangelands/technical-notes-and-fact-sheets/land-soil-vegetation-technical-information
Queensland	Neldner *et al.* (2022)	https://www.qld.gov.au/environment/plants-animals/plants/herbarium/mapping-ecosystems
South Australia	Heard and Channon (1997)	https://cdn.environment.sa.gov.au/environment/docs/vegetation_survey_manual.pdf
Tasmania	Natural and Cultural Heritage Division (2015)	https://nre.tas.gov.au/Documents/Guidelines%20for%20Natural%20Values%20Surveys%20related%20to%20Development%20Proposals.pdf
Victoria	DSE (2004)	https://www.environment.vic.gov.au/__data/assets/pdf_file/0016/91150/Vegetation-Quality-Assessment-Manual-Version-1.3.pdf https://www.environment.vic.gov.au/native-vegetation/native-vegetation
Western Australia	EPA (2016)	https://www.epa.wa.gov.au/sites/default/files/Policies_and_Guidance/EPA%20Technical%20Guidance%20-%20Flora%20and%20Vegetation%20survey_Dec13.pdf

GENERAL PRINCIPLES

The decision to collect vegetation field attributes ultimately depends on the purpose of the survey and should be well documented. All necessary site characteristics (as outlined in earlier chapters of the Field Handbook) should also be recorded. The field standards for vegetation promote the collection of core attributes and when required, additional attributes. It is recommended that quantitative data be collected as opposed to categorical data, whenever possible. Standardised and quantitative assessment ensures the data are appropriate for multiple purposes, such as vegetation and land resource mapping, habitat recognition, monitoring, vegetation classification, threatened species surveys and species distribution modelling. It enables the data to be re-used endless times, representing a significant quality and efficiency gain (collect once, use many times).

A good example is the application of vegetation plot-based data for floristic vegetation classification. For higher levels of classification using large datasets, it may be appropriate to use presence/absence data. However, more detailed classifications may require quantitative data (i.e. floristics with species cover) to inform the analysis and generate the most dominant species representing a vegetation type.

PURPOSE

Here we describe how this chapter relates to several other documents for consistent field sampling, mapping, describing and classifying Australian vegetation types.

1. *Field standards* (this Field Handbook): recording vegetation attributes in the field.
2. *Guidelines* (Thackway *et al.* 2008): guidelines and methods for mapping vegetation communities.
3. *National Vegetation Information System Framework* (*Australian Vegetation Attribute Manual*, NVIS Technical Working Group 2017): concepts and procedures of the NVIS classification scheme and how to put together a vegetation community description using the six-level vegetation hierarchy.

The purpose of these field standards is to outline the attributes recorded in the field to be able to describe and classify vegetation, whether it be at a site level or at the community level. The attributes are not limited for a specific purpose, classification system or typology. The intention is a standard approach to record vegetation attributes enabling reuse of the data for various purposes. Other chapters in the Field Handbook should be referred to for recording site characteristics.

The *Guidelines* (Thackway *et al.* 2008) provide methods for the capture, interpretation and management of vegetation data and information, specifically for vegetation mapping purposes. The *Guidelines* meet the requirements of NVIS (NLWRA 2001; ESCAVI 2003; NVIS Technical Working Group 2017) and contains the national classification scheme. The *Guidelines* aim to assist vegetation scientists to survey, classify and map vegetation types at the association and sub association level of detail (NVIS Technical Working Group 2017). They cover aspects on survey design and planning, field attributes, data analysis and generation of final outputs. The *Guidelines* also provide guidance on timing, sample site location and sample detail for vegetation and ancillary information.

The *NVIS Framework* provides guidance on how to document vegetation data using a consistent set of structural and floristic attributes that are collected, classified and mapped using state, territory and non-government mapping systems. The NVIS *Australian Vegetation Attribute Manual* contains the required tables to generate descriptions at six levels of the NVIS vegetation hierarchy. Specifications for the NVIS vegetation hierarchy are provided in that manual, with examples and how to apply dominance and co-dominance notations.

OVERVIEW OF VEGETATION CLASSIFICATION

Classification systems are often hierarchical, meaning that vegetation types are organised in hierarchical classification levels and qualified using ranks (e.g. association or alliance). In addition, hierarchical systems usually include nested relationships between vegetation types of different ranks (De Cáceres *et al.* 2015). They are typically reliant on vegetation plot data and multivariate analysis.

Vegetation classification has developed from a descriptive into a numerically intensive discipline with an extensive literature and complex protocols and methods (Mucina 1997; De Cáceres *et al.* 2015). Data collection includes decisions regarding survey type (random, stratified, preferential; De Cáceres *et al.* 2015), data type (presence-absence, frequency, cover, cover-abundance) and plot size (Neldner and Butler 2008; Dengler *et al.* 2009; Patykowski *et al.* 2021). Analysis decisions extend from data transformation (Lewis *et al.* 2021) and the type of association metric and clustering algorithm used (e.g. hierarchical *v.* non-hierarchical clustering), to whether environmental data are integral or secondary to the classification (Luxton *et al.* 2021).

Australian vegetation classification has been reviewed in several places, most recently by Luxton *et al.* (2021). Australia has several state-wide classifications of vegetation at varying scales, with each system having different purposes, advantages, limitations and generally with little international context (Sun *et al.* 1997; Beadle 1981; Carnahan 1986; Specht *et al.* 1995; NLWRA 2001; Beard *et al.* 2013; Keith 2017; Gellie *et al.* 2018; Neldner *et al.* 2022). The most recent continental scale classification system is NVIS, which is a structural-physiognomic classification system designed for vegetation mapping.

The field attributes outlined in this chapter are applicable for use in any classification system, examples of which include NVIS (bottom-up approach), and the International Vegetation Classification (IVC; top-down approach). Thus, whether a classification system is physiognomic or floristically dominant, all core attributes that should be recorded in the field are applicable.

National Vegetation Information System Framework

NVIS is a classification scheme and typology, developed for the specific purpose of vegetation mapping in Australia. However, the NVIS framework can also be used at the site level for describing vegetation types. It is a six-level hierarchical classification with the broadest level (1 – class) identified by the dominant growth form, to the detailed level (6 – sub-association) describing up to five dominant species and structural formations for a maximum of nine substrata identified at a site, or across vegetation units (Table 14; Thackway *et al.* 2008; NVIS Technical Working Group 2017).

Table 14 Six hierarchical levels of the National Vegetation Information System Vegetation Hierarchy

Level	NVIS Vegetation Hierarchy	Description
1	Class	Dominant growth form for the structurally dominant stratum (dominant growth form).
2	Structural formation	Structural formation for the structurally dominant stratum (cover, average height, dominant growth form).
3	Broad floristic formation	Dominant genus (or genera) and structural formation for the structurally dominant stratum (dominant genus/genera, cover, average height, dominant growth form).
4	Sub-formation	Dominant genus (or genera) and structural formation for up to three strata (dominant genus/genera, cover, average height, dominant growth form).
5	Association	Up to three dominant species and structural formations for up to three strata (three dominant species, cover, average height, dominant growth form).
6	Sub-association	Up to five dominant species and structural formations to a maximum of eight substrata (five dominant species, cover, average height, dominant growth form).

Adapted from NVIS Technical Working Group (2017).

REMOTE SENSING

This chapter focuses on the standards for recording vegetation attributes in the field. While remote sensing technologies are increasingly used, they are out of scope for this chapter. However, the development of remote sensing technology has opened numerous possibilities to retrieve information on vegetation structural and floristic attributes. Remote sensing can be seen as a tool or method for capturing vegetation data remotely, although the core field attributes do not change (i.e. strata, cover, height, growth form, floristics).

With the advent of drones with optical sensors, LiDAR instruments and terrestrial laser scanning, vegetation data can now be captured remotely at very high spatial resolutions, equivalent to (or at higher levels of precision than) field site data. Optical sensors and LiDAR instruments can be used to capture vegetation cover and height attributes. A terrestrial laser scanner can also capture height and cover attributes as well as other tree volumetrics. Such remote data can then be combined with other field data and satellite remote sensing to support broad-scale analyses (Liang and Wang 2020). For example, LiDAR data can be used to quantify vegetation structure, such as the area of crowns with gaps, understorey vegetation and heights (Calders *et al.* 2020). Multi-spectral remote data can be used to determine *fractional vegetation cover* (photosynthetic vegetation cover, non-photosynthetic vegetation cover, and bare ground) over a site (Muir *et al.* 2011).

High resolution remote sensing methods are not the only way to determine fractional vegetation cover in the field for use with satellite and airborne imagery. The *AUSPLOTS Rangelands Survey Protocols Manual* (White *et al.* 2012) describes measurement of downwards and upwards presence of vegetation at 1010 point intercepts across 1 hectare. From this data, vegetation structure and cover measures relevant to the ground pixel of satellite platforms can be derived. Therefore, it is important to note that if collecting field metrics to scale-up to satellite imagery, it must be fit for purpose as per widely used methods including Muir *et al.* (2011), Sparrow *et al.* (2020) and Laws *et al.* (2023a,b).

FIELD ATTRIBUTES

Vegetation is characterised and classified on the basis of *structure* (the vertical and horizontal distribution of vegetation: its growth form, height, cover and strata) and *floristics* (the dominant genera or species in various strata and characteristic plant species). The characterisation of vegetation types is based on quantitative vegetation plot data (Faber-Langendoen *et al.* 2014). Therefore, emphasis should be placed on collecting primary data (absolute values) in the field, rather than placing the data into pre-determined classes (categorical data) as this extends the usefulness of the data for a variety of purposes (Sun *et al.* 1997). It is essential the core attributes are recorded as outlined below, and a simple workflow followed (Figure 6) in order to recognise and record the required field attributes. By following the standards outlined in this chapter, the data collected can easily be converted to any required category as part of a pre-processing step prior to data analysis. Additional attributes may be recorded, although their collection is highly dependent on the purpose of the survey.

Core vegetation field attributes include:

* strata (layering of vegetation, defined by growth form and height characteristics)
* structure (height, cover)
* growth form (physiognomy)
* floristics (species composition and structure)

1. Decide vegetation type	2. Recognise strata (or growth form & height characteristics)	3. Record structure and floristics
• native (non-rainforest) • tropical/subtropical rainforest • temperate rainforest • non-native vegetation • wetland • mixture • plantation • no vegetation	• emergents • upper stratum i.e dominant • mid-stratum (if present) • ground stratum (if present) • dominant growth form for each stratum • cover and cover type for each stratum • heights and height type for each stratum	• floristics (full, dominant species or stratum specific) • species cover • species height • additional attributes (if required)

Adapted from Hnatiuk *et al.* (2009)

Figure 6   Core vegetation field attributes required to describe and classify native and non-native vegetation types and the workflow required in the field.

Rainforests are covered later in this chapter. Tropical (including subtropical) rainforests follow the classification system of Webb (1978), and oceanic cool temperate rainforests[6] follow the classification systems of Jarman *et al.* (1991) and Reid *et al.* (1999). For all other native vegetation types, the required field attributes are covered in this section. For non-native vegetation, the same principles apply to that of native vegetation. Application of field-based data into the NVIS classification framework is covered briefly later in this chapter. For greater detail, refer to the *Australian Vegetation Attribute Manual* (NVIS Technical Working Group 2017).

The terms *site* and *plot* are used throughout this chapter. A site is defined as a small area of land considered to represent a homogenous vegetation type that contains uniform floristics and structural composition of consistent patterning and density. A plot is a fixed area being sampled to represent a site (e.g. within a 20 × 20 m plot), unless it is a plotless method employed to capture *basal area* using a basal wedge. When upscaling field data to satellite imagery, the site and plot size need to be appropriate for the ground pixel size of the sensor.

Recognising strata

A *stratum* is an easily observed layer of foliage and branches of a measurable *height* (including average and range (minimum and maximum)). Vegetation can have one or more strata. A single stratum may extend from the top of the canopy to near ground level and can be referred to as a continuum where it is difficult to recognise distinct strata. *Strata* are defined by *growth form* and *height* characteristics.

Three major strata are defined as *upper* (**U**), *mid* (**M**) and *ground* (**G**) (NVIS Technical Working Group 2017). Depending on the complexity of the vegetation, subdivisions of the three main strata can be recognised and are referred to as substrata. Substrata occur when a major stratum is composed of two or more distinct layers easily quantified by height as

---

6   Referred to as Tasmanian rainforests in previous editions

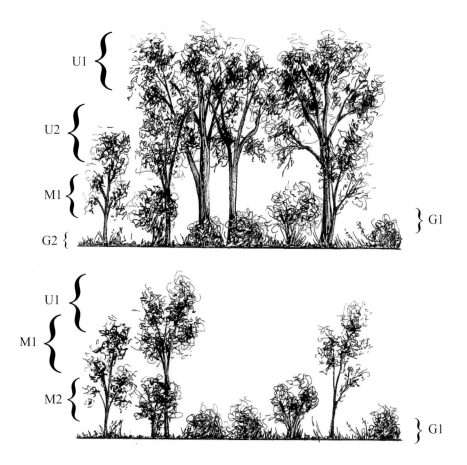

Figure 7 Examples of applying NVIS vegetation stratum and substratum codes to two structurally different vegetation types. Adapted from NVIS Technical Working Group (2017).

Table 15 Growth form and height – a surrogate for recognising strata

Stratum	Stratum (NVIS equivalent)	Growth form	Height range	Notes
E: emergent	U1	Tree	≤½ mean ht of E	
T1: canopy	U1 (if emergent absent)	Tree	≤½ mean ht of T1	
T2: sub-canopy	U2	Tree	≤½ mean ht of T2	
T3: low tree layer	U3	Tree	Trees <T2 layer	
S1: tallest shrub layer	M1	Shrub (low trees)	≤8 m	
S2: lower shrub layer	M2	Shrub	Distinct layer below S1	Infrequently recognised
G: ground layer	G1	Graminoids/forbs/ sprawling vines, seedlings	0 m ≤ G ≤ 2 m; usually	Height measured to top of main leaf biomass

Adapted from Neldner *et al.* (2022).

Table 16 NVIS strata and substrata codes, descriptions, applicable growth forms and heights

NVIS stratum code	NVIS substratum code	Description	Traditional stratum name	Growth forms*	Height classes**	Not permitted
U	U1	Tallest tree substratum. For forests and woodlands this will generally be the dominant stratum, except where the substratum cover is <5% and hence considered 'emergent'. For a continuum (e.g. no distinct or discernible layering in the vegetation), the tallest stratum becomes the defining substratum.	Upper, tree Overstorey/ canopy (if only one tree layer occurs it is coded U1)	Trees, tree mallees, palms, vines (mallee shrubs) Also: epiphytes, lichens	8, 7, 6 (5)	Grasses and shrubs, low mallee shrubs
	U2	Sub-canopy layer, second tree layer				
	U3	Sub-canopy layer, third tree layer				
M	M1	Tallest shrub layer	Mid, shrub (if only one mid layer occurs it is coded M1)	Shrubs, low trees, mallee shrubs, vines, (low shrubs, tall grasses, tall forbs, tall sedges) grass trees, tree ferns, cycads, palms. Also: epiphytes, lichens	(6) 5, 4, 3	Mid and low grasses, sedges, rushes and forbs Mid and tall trees/palms
	M2	Next shrub layer				
	M3	Third shrub layer				
G	G1	Tallest ground species	Lower, ground (if only one ground layer occurs it is coded G1)	Grasses, forbs, sedges, rushes, vines, lichens, epiphytes, low shrubs, ferns, bryophytes, cycads, grass trees, aquatics, seagrasses.	(4) 3, 2, 1	Trees, tree-mallees, palms
	G2	Ground				
	G3	Substrate surface		Bryophytes, lichens, lower plants	1	

Adapted from NVIS Technical Working Group (2017).
* See Table 17 for further details.
** See Table 19 for further details.

illustrated in Figure 7. If strata are not recognised and documented in the field, growth form and height characteristics can be used as a surrogate (Table 15).

Table 16 provides NVIS stratum and substratum codes, descriptions, applicable growth forms and heights. The NVIS classification framework defines up to nine strata/substrata to accommodate Australia's structurally complex vegetation types. This chapter recommends the adoption of the NVIS vegetation strata and substratum definitions to ensure field data from across the continent are comparable and reusable.

Emergents

The tallest plants in some vegetation communities are so sparse that they no longer form the dominant or most significant layer. For example, widely scattered eucalypts or acacias may rise above lower shrubs or grasses in semi-arid regions. The tallest stratum is then classified as an 'emergent layer' and the dominant layer on which the vegetation will be classified is usually the next tallest layer (Figure 8).

For example, using the NVIS classification framework, U1 is the tallest tree substratum (Table 16). For forests and woodlands, this will generally be the dominant stratum, except where the substratum cover is <5% and hence considered 'emergent'. For a continuum (e.g. no distinct or discernible layering in the vegetation), the tallest stratum becomes the defining substratum (NVIS Technical Working Group 2017).

As a guideline, emergents are recognised if their foliage cover is less than 5% of the crown cover of the dominant stratum. Care is needed in classifying tall plants as emergents, and the 5% guideline can be expected to vary depending on vegetation type and season – especially if the ground layer is the dominant stratum. In some borderline cases, taller plants can occur over a lower stratum that elsewhere forms the dominant stratum of a well-known vegetation type. In this situation it is acceptable to continue to call the tallest stratum emergents where

Figure 8   An emergent layer for a hypothetical site with four strata. Adapted from Hnatiuk *et al.* (2009).

it is considered unhelpful to create a new vegetation type solely based on the unusual, slightly higher cover of the tallest plants.

Complex canopies due to regrowth

Vegetation that has been disturbed or is still recovering from certain kinds of disturbance can produce complex canopies. For example, where a canopy has been reduced – but not totally removed – by clearing, ringbarking or poisoning, two or more cohorts of canopy species may occur. When of clearly different ages, the cohorts are also likely to differ in height, making the description of the canopy difficult.

The methods already described for defining dominant and emergent strata should be applied to these situations. This type of vegetation can be further characterised by recording 'uneven age' when assessing *growth stage* (refer to 'Growth stage' section in this chapter). The different cohorts should not be amalgamated unless they are too similar in structure to be distinguished consistently. Arbitrary height boundaries should not be used to separate them as far as is practicable.

Structural attributes

Vegetation usually consists of a mixture of *growth forms* (such as trees, shrubs, grasses) of varying *height* (strata, layers or continua) and *cover* (as a percentage). These three features (growth form, height and cover) account for most of the appearance of the vegetation and are used to classify its *structure*. These structural attributes are required to be recorded in the field to derive a structural formation (e.g. level 2 of the NVIS vegetation hierarchy) – for example, using the NVIS classification framework to create a field-based vegetation type description.

Growth form

The term *growth form* is used in a broad sense to describe the form or shape of individual plants (e.g. tree, shrub) or Australian broad floristic land cover types (e.g. native vegetation such as mallee and chenopod shrub, or non-native vegetation such as wheat crops and orchards). The standards in this chapter recommend the use of the NVIS growth form terminology. The application of a consistent vocabulary ensures data collected by multiple organisations (including consultants and land managers) are comparable, irrespective of the classification system chosen. Table 17 provides the NVIS growth form types and descriptions and encourages the adoption of these when describing and classifying Australian vegetation types.

The term *growth stage* is often confused with growth form (also sometimes referred to as lifeform). Growth stage refers to the phenological life cycle of a plant where the broad stages in this context include regeneration, mixed age, mature phase and senescent phase for vegetation communities.

Cover

Cover is the area of the ground surface covered by vegetation and is expressed as a percentage of area. Vegetation cover is broadly measured in two ways using field methods or remote sensing methods. Here we discuss field measures only, at a sample site. Definitions that clearly describe what is being captured will improve the consistent application and interpretation of vegetation cover measures. The three commonly used field measures of cover are *crown* (or *canopy*) *cover*, *foliage cover* and *foliage projective cover*. Each gives different values for the same vegetation cover.

Table 17 NVIS growth forms and definitions

	Growth form	Definition
T	Tree	Woody plants, more than 2 m tall with a single stem or branches well above the base.
M	Tree mallee	Woody perennial plant usually of the genus *Eucalyptus*. Multi-stemmed with fewer than five trunks of which at least three exceed 100 mm diameter at breast height (1.3 m). Usually 8 m or more.
S	Shrub	Woody plants multi-stemmed at the base (or within 200 mm from ground level), or if single stemmed, less than 2 m.
Y	Mallee shrub	Commonly less than 8 m tall, usually with five or more trunks, of which at least three of the largest do not exceed 100 mm diameter at breast height (1.3 m).
Z	Heath shrub	Shrub usually less than 2 m, with sclerophyllous leaves having high fibre:protein ratios and with an area of nanophyll or smaller (less than 225 mm^2). Often a member of one the following families: Ericaceae, Myrtaceae, Fabaceae or Proteaceae. Commonly occurs on nutrient-poor substrates.
C	Chenopod shrub	Single or multi-stemmed, semi-succulent shrub of the family Chenopodiaceae (and some chenopods now in the family Amaranthaceae) exhibiting drought and salt tolerance.
U	Samphire shrub	Genera with articulate branches, fleshy stems and reduced flowers, succulent chenopods. Includes the genus *Tecticornia* (and previously *Halosarcia*, *Pachycornia*, *Sarcocornia*, *Sclerostegia* and *Tegicornia* (incorporated into *Tecticornia*)). Also the genus *Suaeda*.
G	Tussock grass	Grass forming discrete but open tussocks usually with distinct individual shoots, or if not, then forming a hummock. These are the common agricultural grasses.
H	Hummock grass	Coarse xeromorphic grass (genus *Triodia*) with a mound-like form often dead in the middle.
W	Other grass	Member of the family Poaceae, but having neither a distinctive tussock nor hummock appearance. An example species is *Cynodon dactylon*.
V	Sedge	Herbaceous, usually perennial erect plant generally with a tufted habit and of the families Cyperaceae (true sedges) or Restionaceae (node sedges).
R	Rush	Herbaceous, usually perennial erect monocot that is neither a grass nor a sedge. For the purposes of NVIS, rushes include the monocotyledon families Juncaceae, Typhaceae, Liliaceae, Iridaceae, Xyridaceae and the genus *Lomandra* (i.e. 'graminoid' or grass-like genera).
F	Forb	Herbaceous or slightly woody, annual or sometimes perennial plant (usually a dicotyledon).
D	Tree fern	Characterised by large and usually branched leaves (fronds), arborescent and terrestrial; spores in sporangia on the leaves.
E	Fern	Ferns and fern allies (except tree fern). Characterised by large and usually branched leaves (fronds), herbaceous and terrestrial to aquatic; spores in sporangia on the leaves.
B	Bryophyte	Mosses and liverworts. Mosses are small plants usually with a slender leaf-bearing stem with no true vascular tissue. Liverworts are often moss-like in appearance or consisting of a flat, ribbon-like green thallus.
N	Lichen	Composite plant consisting of a fungus living symbiotically with algae; without true roots, stems or leaves.
K	Epiphyte	Epiphytes, mistletoes and parasites. Plant with roots attached to the aerial portions of other plants. Can often be another growth form, such as fern or forb.
L	Vine	Climbing, twining, winding or sprawling plants usually with a woody stem.

Table 17 (cont.)

	Growth form	Definition
P	Palm	Palms and other arborescent monocotyledons. Members of the Arecaceae family or the genus *Pandanus*. *Pandanus* is often multi-stemmed.
X	Grass tree	Australian grass trees. Members of the Xanthorrhoeaceae family.
A	Cycad	Members of the families Cycadaceae and Zamiaceae.
J	Seagrass	Flowering angiosperms forming sparse to dense mats of material at the subtidal level and down to 30 m below MSL. Occasionally exposed.
Q	Aquatic	Plant growing in an inland waterway or wetland with the majority of its biomass under water for most of the year. Fresh, saline or brackish water.
O	Lower plant	Algae, fungus.

Adapted from NVIS Technical Working Group (2017).

Table 18 Recommended cover types, with definitions

	Cover type	Definition
1N	Crown or canopy cover (CC)	The percentage of the sample plot within the vertical projection of the periphery of the crowns. Crowns are treated as opaque (Walker and Hopkins 1990).
2N	Foliage cover (FC)	The percentage of the sample plot occupied by the vertical projection of foliage and branches (if woody) (Walker and Hopkins 1990). For ground vegetation, it is measured using line-intercept methods. It will, to some degree, take into account the thickness of a clump of grass (Walker and Hopkins 1990).
3N	Percentage cover	The percentage of a strictly defined plot area, covered by vegetation, generally applicable for the ground vegetation that has been estimated rather than measured using line-intercept methods. It does not necessarily take into account thickness of a clump of grass.
4N	Foliage projective cover (FPC)	The percentage of the sample plot occupied by the vertical projection of foliage only (Walker and Hopkins 1990).

Adapted from NVIS Technical Working Group (2017).

Crown cover (Walker and Hopkins 1990) is the percentage of the sample site within the vertical projection of the periphery of crowns, with crowns considered to be opaque. This is also the generic definition of canopy cover or plant cover.

Foliage cover (Carnahan 1977; Walker and Hopkins 1990) is the percentage of the sample site occupied by the vertical projection of foliage and woody branches. Foliage cover can be estimated from crown cover.

Foliage projective cover (FPC), also referred to as projective foliage cover (PFC) as defined by (Specht *et al.* 1974), is the percentage of the sample site occupied by the vertical projection of foliage only and is usually expressed as the percentage of ground covered by foliage. It can be separated according to vegetation strata, particularly overstorey or understorey.

FPC and foliage cover are sensitive to season and drought because foliage may change greatly depending on water availability. This variation is not usually considered in vegetation classification. Therefore, irrespective of the application, it is essential that the cover type is recorded for each stratum, and ideally should be consistent across all strata. Another cover type worth noting is *percentage cover*. This is an estimate that is generally applicable to ground cover within a defined plot area. Table 18 provides cover types and descriptions adapted from the NVIS classification framework. These are the recommended cover types to be used when describing and classifying vegetation types in Australia.

Point intercept

Both crown cover and FPC may be collected objectively and simultaneously when using the point-intercept method (Figure 9). The point-intercept method is commonly established in a grid that is aligned to the plot defined boundaries (Laws *et al.* 2023a). To discriminate between cover types (crown cover and FPC), the concept of '*in-canopy sky*' can be employed. The canopy perimeter, from a vertical perspective, is typically described by the extent of the outer layer of leaves of an individual tree or shrub (White *et al.* 2012). Within the canopy, openness or density of foliage varies depending on species, leaf shape and deciduousness. In-canopy sky is applied when using the point-intercept method when under a canopy and foliage or branches are not intercepted using a densitometer (Figure 9). This concept is taken from field techniques implemented for remote sensing field validation (Muir *et al.* 2011). It is worth noting that this methodology can be similarly applied to lower vegetation strata (e.g. the ground layer) to obtain accurate measures of vegetation cover for growth forms other than trees (Laws *et al.* 2023a).

Ground cover is routinely collected in many surveys where vegetation is being assessed. The point-intercept method is suitable for collecting ground cover (or sometimes referred to as substrate) by recording the nature of the substrate which is intercepted. Depending on the purpose of the survey the categories for substrate may include, but are not limited to, bare, cryptogam, gravel, litter, outcrop and rock (Laws *et al.* 2023a). This method can be applied simultaneously with recording lower vegetation strata to obtain accurate measures of vegetation cover for growth forms other than trees.

Ground cover can also be estimated by measuring the distance covered by the vertical projection of the leaves and woody branches onto a tape measure and expressing this as a percentage of the total length. This is referred to as a line-intercept method.

Point intercepts to record

0	nil-outside canopy
1	nil-outside canopy
2	nil-outside canopy
3	nil-outside canopy
4	**canopy hit**
5	nil-outside canopy
6	nil-outside canopy
7	**in canopy sky**
8	nil-outside canopy
9	nil-outside canopy
10	**canopy hit**
11	**canopy hit**
12	nil-outside canopy
13	**in canopy sky**
14	nil-outside canopy
15	nil-outside canopy

Figure 9 Example of point intercept 'upwards' determination of canopy. Source: Laws *et al.* 2023b: cover module.

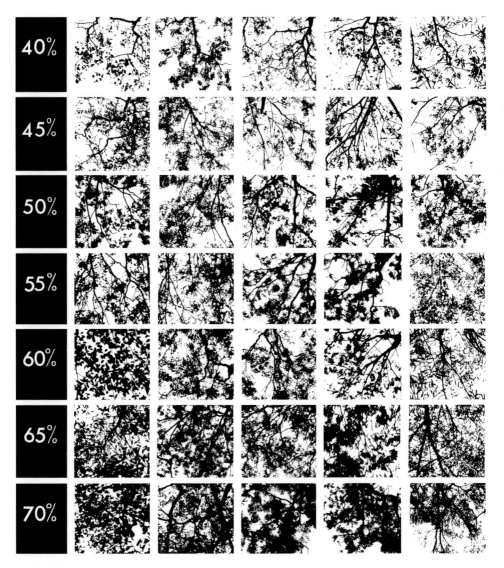

Figure 10 Crown types. Estimate the openness of individual tree or shrub crowns by matching the crown with a photograph. The rows show similar crown types for different leaf size (large to small, left to right). *Acacia* phyllodes are in the right-hand column. Most Australian woody plants are in the range 40–70%. Source: Hnatiuk *et al.* (2009) and Walker and Hopkins (1990).

Crown separation ratio

Crown separation ratio (CSR) provides a rapid approximation of vegetation cover. Crown cover percentage can be estimated using CSR developed by Walker *et al.* (1988) and Penridge and Walker (1988). CSR is the ratio of the mean gap between crowns and the mean crown width, that is:

$$CSR = \text{mean gap between crowns/mean crown width}$$

CSR is not reliable if crowns deviate significantly from circular or slightly oval (e.g. in forests with a significant cover of *Corymbia* that have very irregular and interlocking shapes). The CSR method is available in Walker and Hopkins (1990) and Hnatiuk *et al.* 2009).

A rapid estimation of cover can be assessed by matching the crown in the field with a photograph provided in Figure 10. Note that this is a very rapid method and is not regarded best practice.

Height

In the field, actual height measurements should be made where possible, rather than estimates or height classes. Inaccuracy in measurements increases as crown closure and height increase. There are several height types, although for general vegetation mapping applications average or median height is the most common, where the bulk of the vegetative material or canopy layer falls within any stratum (Figure 11). For applications such as forestry, top height is used as it provides the general height of the top of the tallest canopy layer (i.e. tallest tree) which may not necessarily be the dominant canopy layer. Similarly, layer height measures to where the bulk of the canopy is. For all height types, canopy depth can be characterised by the minimum and maximum height values.

Irrespective of the application, it is essential that the height type is documented for each stratum (i.e. average or median height is generally used for most applications). Table 19 provides the recommended NVIS height types and encourages the adoption of these in order to describe and classify vegetation types in Australia consistently.

Height can be measured using measuring tapes or poles for low vegetation. Clinometers, laser or sonic ranging instruments, and visual sighting instruments can be used for tall vegetation (Brack 1998; Abed and Stephens 2003). Use of clinometers and range finders for vegetation measurement is discussed in White *et al.* (2012) and Laws *et al.* (2023a).

Record the height from the ground to the highest part of the plant above ground. Where the height of flower stalks (e.g. in grasses, grass trees) or leaves (e.g. in palms, cycads, grass

Figure 11   Height types. Adapted from ESCAVI (2003).

Table 19   Recommended height types, definitions, applicable strata and structural formations

Code	Height type	Stratum	Structural formation	Definition
NA	Average height (general vegetation mapping)	Any	Forest, woodlands, shrublands, grasslands	Average height of the stratum where the bulk of the vegetative material falls within any particular stratum. This may be the measurement of a number of individuals, which fall within the range of the recognised stratum. The mean value becomes essentially a measure of the midpoint of the canopy depth. The minimum and maximum values define the depth of the canopy or layer.
NV	Layer height (general vegetation mapping)	Any	Forest, woodlands, shrublands, grasslands	Layer height of the top stratum (top of the canopy or the top of the bulk of the vegetative material making up the stratum) that may be present. The minimum and maximum values of this will not give any indication of canopy depth if applied to taller strata.
NT	Top height	U1, U2, M1	Forests woodlands, shrublands	General height of the top of the tallest canopy layer (i.e. tallest tree) which may not necessarily be the dominant canopy layer. The minimum and maximum values will not give any indication of canopy depth. This height category may indicate U1 as 'emergent' layer and U2 as the dominant layer.

Source: NVIS Technical Working Group (2017).

Table 20   Height classes in relation to growth form as described in the NVIS framework

Height		Growth form				
	Height (m)	Tree, palm (single-stemmed), vine (*U)	Tree mallee, mallee shrub (*U)	Shrub, heath shrub, chenopod shrub, ferns, samphire shrub, cycad, tree fern, grass tree, palm (multi-stemmed) (*M & *G)	Tussock grass, hummock grass, other grass, sedge, rush, forbs, vine (*G)	Bryophyte, lichen, seagrass, aquatic (*G)
8	>30	tall	–	–	–	–
7	10–30	mid	tall	–	–	–
6	<10	low	mid	–	–	–
5	<3	–	low	–	–	–
4	>2	–	–	tall	tall	–
3	1–2	–	–	mid	tall	–
2	0.5–1	–	–	low	mid	tall
1	<0.5	–	–	low	low	low

* NVIS stratum code.
Source: NVIS Technical Working Group (2017).

trees, tree ferns) add significantly to plant height and contribute significantly to a stratum, then record two measurements: total height from ground level to the top of the highest part of the plant, and height from ground level to the top of the leaves (e.g. *Xanthorrhoea johnsonii* 2.5 m/1.3 m; *Sorghum intrans* 1.9 m/1.3 m). This provides an accurate record and allows various uses in analysis.

Hnatiuk *et al.* (2009) provided a list of plain descriptors for height classes, in relation to woody and non-woody plants, that remain useful. The NVIS framework height classes in relation to growth form are listed in Table 20.

Floristics

Floristics is the list of plant species occurring at a sample site or within a plot. Ideally, all species present in a plot should be listed in order of tallest to lowest stratum/substratum, as this extends the usefulness of the data. However, the completeness of a full floristic species list will often depend on the degree of botanical expertise/training, familiarity with the floristics of a geographic area, and the time of sampling. Ideally, full scientific names will be used to avoid ambiguities and to make it easier to combine datasets. If using *ad hoc* species names (i.e. field name), ensure that voucher specimens are collected, and records are updated with current and accepted scientific names, as per the process described in White *et al.* (2012). Field manuals for each jurisdiction (Table 13) will also contain guidance for the documentation of field names and collection of specimens.

Structural information is also recorded at the species (or in some cases infraspecific) level. For each species recorded within a plot, the strata in which it occurs should be recognised, and if possible, its growth form, cover (as crown cover or FPC) and height (including average or median, and range). This information is critical to determine floristic composition, abundance, and is increasingly useful data in numerical analysis for floristic classification.

Species nomenclature and taxonomy is a somewhat dynamic space. Consequently, the recording of floristics in the field should follow the recommended nomenclature and taxonomy for the purpose the work is being conducted and the region you are working in.

Multiple avenues exist at the national and local level, in a variety of formats, to access taxonomic data. Commonly used digital lists are provided in Table 21. The National Species List (NSL) comprises both the Australian Plant Census (APC) and the Australian Plant Name Index (APNI). The APC is a list of accepted scientific names for the Australian vascular flora, both native and introduced, and includes synonyms and misapplications for these names. The APC covers all published scientific plant names in the Australian context in the taxonomic literature, but excludes taxa known only from cultivation in Australia. The data comprises names, bibliographic information, and taxonomic concepts for plants that are either native to or naturalised in Australia. APNI is a tool for the botanical community that deals with plant names and their usage in the scientific literature, whether as a current name or synonym. APNI does not recommend any particular taxonomy or nomenclature, which is where the APC should be used for current names and is endorsed by the Council of Heads of Australasian Herbaria (CHAH).

Good field floristic work is based on the following practices:

- Ensure that appropriate collecting permits and/or permissions are obtained before collecting.
- Know what constitutes an adequate specimen for the various types of plants you will encounter.
- Know what rare flora may be encountered, how to identify it and what to do if any are found.
- Record the basic information for voucher specimens: plant name, location where collected (geo-coordinates, distance/direction from known geographic feature), date, collector and collector's number, habitat (soil, vegetation type, etc.), plant height, phenological state (flowering, fruiting, leafing, dormant, colours of plant parts, etc.).

Table 21   National Species List services, electronic floras

National Species List services	
Vascular plants	
Australian Plant Census	https://biodiversity.org.au/nsl/services/search/taxonomy
Australian Plant Name Index	https://biodiversity.org.au/nsl/services/search/names
Lichens	https://lichen.biodiversity.org.au/nsl/services/
AusMoss	https://moss.biodiversity.org.au/nsl/services/
Fungi	https://fungi.biodiversity.org.au/nsl/services/
Electronic floras and checklists	
Flora of Australia Online	https://profiles.ala.org.au/opus/foa
VicFlora (Victoria)	https://vicflora.rbg.vic.gov.au/
FloraBase (Western Australia)	https://florabase.dbca.wa.gov.au/
PlantNet (New South Wales)	https://plantnet.rbgsyd.nsw.gov.au/
FloraNT (Northern Territory)	http://eflora.nt.gov.au/
eFlora SA (South Australia)	http://www.flora.sa.gov.au/
Flora of Tasmania	https://flora.tmag.tas.gov.au/
Flora Census (Queensland)	https://www.data.qld.gov.au/dataset/census-of-the-queensland-flora-and-fungi-2022
EUCLID – Eucalypts of Australia (including *Corymbia* and *Angophora*)	https://apps.lucidcentral.org/euclid/text/intro/index.html
WATTLE – Acacias of Australia	https://apps.lucidcentral.org/wattle/text/intro/index.html

- Tag each specimen, recording the collector's name/initials and field number which should be unique to the collector or the project and which will also be recorded on the field data sheets or an app.
- Preserve the plants by pressing and drying in a plant press. Some types of plants may need special treatment (e.g. mosses, lichens, fungi, algae, aquatic plants, succulents, very large plants/leaves).
- If using field names and identification numbers, ensure subsequent updating of records when formal identification is complete.
- See Western Australian Herbarium: https://www.dbca.wa.gov.au/science/research-tools-and-repositories/wa-herbarium
- See Queensland Herbarium: https://www.qld.gov.au/environment/plants-animals/plants/herbarium/specimens

Additional attributes

Depending on the purpose of a survey, additional field attributes may be recorded. The collection of these attributes can value-add to the data collected (i.e. core attributes) and enhance data analysis capabilities. For the attributes listed, the most up to date and relevant literature is provided to help guide the collection of this data, and to ensure standards and current methods are being referred to.

Basal area

Basal area (BA) is the sum of the cross-sectional area of trees (trees, tall shrubs and/or mallee greater than 2 m in height), including bark, at breast height (1.3 m above ground level) and expressed as metres squared per hectare (m²/ha). Stand BA is the sum of the tree BAs in a defined site or plot. It is a widely used forestry inventory metric to calculate biomass and used in carbon accounting. It can also be adapted to estimate woody species dominance and is not influenced by seasonal fluctuations like foliage cover. BA may be measured *directly* in a fixed area plot (plot sampling) by summing the woody stem diameter at breast height (DBH) in metres as measured 1.3 m above the ground (Figure 12). DBH is converted to BA based on the formula for the area of a circle:

$$BA = \pi \times (DBH \div 2)^2$$

The BA per hectare is calculated by adding the BAs (as calculated above) of all trees in a site and dividing by the area of land in which the trees were measured. BA is generally measured for a plot and then scaled to m²/ha to compare forest/woodland productivity and growth rate among multiple sites.

BA may also be estimated *indirectly* by using the angle count method first developed by Bitterlich in 1948 using a Relaskop (point sampling, or plotless method). In contemporary vegetation surveys, BA is a simple and easy attribute to estimate in the field using a handheld angle gauge or basal wedge. Using a basal wedge extends the area sampled beyond the actual plot to provide a good approximation of BA for each tree and shrub species and total BA in the vegetation type surrounding the plot. Plants outside the plot are included to better represent the broader vegetation type (White *et al.* 2012). BA can also be used to calculate biomass based on allometric equations.

The technique can also be adapted to estimate crown cover for low woody vegetation using a Bitterlich gauge (Friedel and Chewings 1988; DPIRD 2020). The BA count can also be used

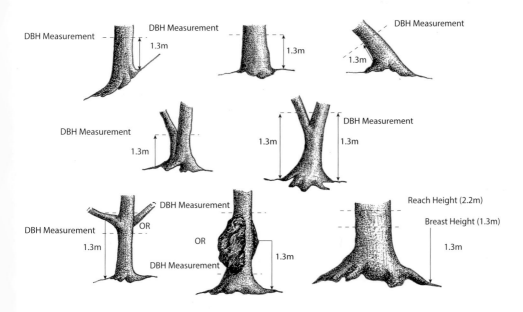

Figure 12 Method for determination of diameter at breast height. Adapted from Wood *et al.* (2015).

to give an unbiased estimate of the dominant woody species at a site. The dominance of woody species at a site can be ascertained by calculating the BA for each species, divided by the total BA occupied by all species at a site.

There are several variations of a BA angle gauge or basal wedge, but essentially each has an aperture (basal factor; BF) through which to view and count tree and shrub stems in a 360-degree radius (Laws *et al.* 2023b). The technique is a variable radius plot measure because the area covered depends on the tree size and basal factor used to count the woody stems. The count may include both live and/or dead trees by individual species. In non-plantation forest situations where the woody species are randomly spaced, the BA count should be repeated at least three times at different sampling points to improve accuracy. To reduce systematic errors, refer to Laws *et al.* (2023b).

For more detail on basal area methodology refer to Laws *et al.* (2023a).

Plant traits and phenology

Plant traits are characteristics or any measurable property capturing aspects of the plant structure or function, including phenological traits. Traits vary in scope from physiological measures of performance (e.g. photosynthetic activity, water-use efficiency, phenological development, disease resistance) to morphological attributes, which can include vegetative characters (i.e. stems, leaves), reproductive characters (i.e. flowers, fruits), quantitative measurements (i.e. number of petals) and qualitative characters (leaf shape, flower colour etc.). Plant traits can link to aspects of ecological variation, and assist in taxonomic revisions and investigations such as evolutionary adaptation. The collection of plant trait data will depend on the purpose of the survey and can either be recorded in the field or taken from physical specimens collected in the field then vouchered in herbaria for downstream analysis.

AusTraits and the Australian data partnerships program have compiled plant trait data collected by researchers from diverse disciplines, including functional plant biology, plant physiology, plant taxonomy and conservation biology. Substantial trait databases now exist for plants (Gallagher *et al.* 2020), with AusTraits containing the Australian plant taxa traits. The database includes 450 traits, representing over 28 000 Australian taxa (Falster *et al.* 2021), and is one of several database initiatives around the world (Boenisch and Kattge 2019; Wigley *et al.* 2020).

Plant phenology is the study of the plant biological cycles which are responsive to changes in the environment. Attributes of phenology can be observed and collected at both the species level within a site, and at the plant community level. Examples of phenological measurements include growth rate, leaf drop/stress tolerance, flowering onset and duration, seed production that responds to environmental changes in temperature, photoperiod, rainfall and nutritional resources.

The collection of plant trait and phenological data will also depend on the purpose of the survey and should ideally be documented in the field. Prior to commencement of a survey, the phenological phase to be measured must be clearly defined to ensure repeat observations are comparable. Repeat observations should be taken at the same place on the same plant/crop each time and include observations of the associated environmental conditions. The frequency will depend on what is being measured and be influenced by changes in environmental conditions. Muir *et al.* (2011) provide guidance as to the attributes to record from a remote sensing perspective, also analogous with growth stages. There are many other purposes for collecting phenology, some of which are in association with flora and fauna surveys (i.e. documenting mass flowering and fruiting events).

Growth stage

The *growth stage* of a plant is its phenological phase in the life cycle, not to be confused with *growth form*. Accurately assessing growth stage can be difficult in unfamiliar vegetation. When a number of species in a vegetation community goes through a phenological phase nearly simultaneously, then it is possible to recognise community growth stages (Cheal 2010). Growth stages are largely a qualitative observation and phases may not be discrete. The growth stages in regenerating vegetation are seen as gradual changes in structure and composition and are sometimes referred to as 'successional' changes. Growth stage is often used to assess regenerating vegetation previously impacted by extreme events such as fire or cyclone. Vegetation appearance can be affected by condition as well as growth stage and it may not be possible to distinguish the effects of age from responses due to stress caused by environmental factors such as pests, diseases or land use. Guidelines for estimating crop growth stages are not covered here but are available from state and territory agricultural departments.

Estimating the age of vegetation, including older growth areas, is possible using modern technology. Archived satellite remote sensing data can be used to estimate age (Lucas *et al.* 2020) and determine time since disturbance impacts that is within the period of available data. Otherwise, for older vegetation, carbon dating and other forestry methods may need to be used.

Table 22 provides an approach for assessing growth stage at the site level for vegetation types, including these growth forms: trees, shrubs, grasses and herbs, and cryptogams. Indicators for the dominant growth form types are described for these growth stages: early regeneration, advanced regeneration, uneven age, mature phase and senescent phase.

The recommended approach to assess growth stages is to walk through the site and surrounding area, looking for signs that indicate the history of the development of the vegetation. Record the growth stage, as outlined in Table 22, as well as the features on which the assessed stage is based.

Where the vegetation is dominated by trees, especially eucalypts in south-eastern and south-western parts of Australia, the signs of ageing are evident and well documented (Jacobs 1955; Eyre *et al.* 2017). Growth stages of trees in sparse vegetation ('woodland' trees) are similar to those in mid-dense and closed vegetation ('forest' trees), but the overall tree-form is shorter and wider (Figure 13). There is little documentation of growth stages for shrubs. Lange and Sparrow (1992) indicate the general cycle of ageing of western myall (*Acacia papyrocarpa*) shrubs in inland Australia, which can be used as a guide for other shrubs.

RAINFORESTS

In Australia, there are patches of rainforest across the tropical north, down the east coast, and in Tasmania. They can occur as extensive contiguous forests, but often occur as scattered patches among sclerophyll vegetation. They are usually easy to distinguish from adjacent forests, which are typically dominated by *Eucalyptus* and other sclerophyllous vegetation. Rainforests tend to have closed canopies that are usually dark green and easily distinguished from the generally

Table 22 Indicators of vegetation community growth stages

	Growth stage	Trees dominant	Shrubs dominant	Grasses and herbs dominant	Cryptogams dominant (mosses and lichens)
1	Early regeneration	Dominated by small, juvenile, dense to very sparse regenerating plants, with or without a few older, widely spaced, emergent plants.	Dominated by small, juvenile, dense to very sparse regenerating plants. A few older, widely spaced emergents may be present.	Small plants and juvenile stages predominate, bare soil or old litter common.	Thin growth of young plants or widely spaced clumps of young plants.
2	Advanced regeneration	Dominated by dense to sparse, well-developed, immature plants. Large emergents can be present with crown cover <5% of the total crown cover. However, if the cover is >5%, classify as 'uneven age'. Trees have well-developed stems (poles). Crowns have small branches. The height is below maximum height for the stand type. Apical dominance still apparent in vigorous trees.	Dominated by dense to sparse, well-developed but not mature plants. If large emergent plants are present, then they occupy <5% crown cover of the dominant stratum; if >5%, classify as 'uneven age'.	Vegetative growth abundant; plants approaching full mature size but reproductive material absent or in early stages only; soil surface largely obscured in average sites.	Cover of plants high for the site; some reproduction may be evident.
3	Uneven age	Mixed size and age classes, usually identified by two or more strata dominated by the same species but with different species regenerating in the understorey of an older canopy.	Mixed size and age classes; usually two or more strata dominated by the same species but includes sites with different species regenerating in the understorey of an older canopy.	A mixture of mature, perennial and immature annual species present on site.	A mixture of mature reproductive plants with immature regeneration.
4	Mature phase	Mature-sized plants, with or without emergent senescent plants. Trees at maximum height for the type and conditions. Crown at full lateral development and limbs thicker in stands. No apical dominance.	Mature-sized plants with or without emergent senescent plants.	Most plants of reproductive age; depending on vegetation type, reproduction evident, or would be if environmental conditions were appropriate (e.g. water availability).	Swards of plants common; plants of mature physiognomy (clump sizes and forms); reproduction common at appropriate times of year or drought/rain cycle; overall health and vigour high.
5	Senescent phase	Dominated by over-mature plants particularly in the dominant stratum; evidence of senescence in many plants, some without obvious links to disturbance. Tree crowns show signs of contracting: dead branches and decreased crown diameter and leaf area. Distorted branches and burls may be common. Dead trees may be present.	Dominated by old plants (thick stems and primary branches, with increasing dead wood or thin and open if species sheds dead branches), particularly in the dominant stratum. Many senescent plants, some without obvious links to disturbance.	In largely annual vegetation, reproduction is complete, and plants are dying or mostly dead; in perennial vegetation, plants have lost vigour, are breaking down; perennial grass base choked with old/brown tiller bases. Litter accumulation may be high.	Clear evidence of the degeneration of plants or clumps; dead older parts of plants may be conspicuous.

Adapted from Hnatiuk *et al.* (2009).

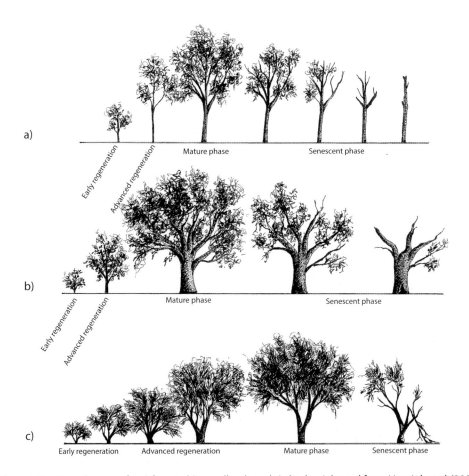

Figure 13 Growth stages for a) forests, b) woodlands and c) shrubs. Adapted from Hnatiuk *et al.* (2009).

greyish and reddish-green canopies of surrounding forests. The 'dry scrubs' of south-east Queensland are closely related to rainforests and are included as such.

The 'dry' rainforests in the Northern Territory, Western Australia and parts of Queensland, as well as the temperate rainforests in south-eastern mainland Australia, are usually described using the standard core attributes. Due to their structural complexity, however, it may not be practical to describe the wet tropical and subtropical rainforests of Australia using the attributes and methods used for other vegetation types. The oceanic cool temperate rainforests of Tasmania can also be complex in structure. These two varieties of rainforests may be sampled using either the standard core attributes, or methods supplemented with extra attributes to fully reflect their additional complexity. The rest of this section deals separately with these two special cases, tropical (including subtropical) and oceanic cool temperate rainforests.

Tropical and subtropical

Australian tropical rainforests are situated above approximately 22° latitude, while subtropical rainforests are between approximately 22° and 33° latitudes. Tropical rainforests are highly diverse (CSIRO 2021) and it is difficult for non-specialist rainforest practitioners to accurately identify the taxa in an area. For this reason, Webb *et al.* (1976) tested the value of structural characteristics in describing and typifying tropical rainforests. This was further developed into a structurally based typification (Webb 1978). Although knowledge of, and expertise in, the floristic taxa in tropical and subtropical rainforests has improved, the structural typification of tropical and subtropical rainforests remains a useful tool for non-specialists (Neldner *et al.* 2022).

The tropical and subtropical rainforests in different jurisdictions are described using:

- New South Wales: the core vegetation attributes of species, growth form, strata, height and cover, and the methods used in that jurisdiction (Sivertsen 2009). NSW has classified subtropical rainforests using a numerical analysis of floristic data and environmental attributes (DPE 2022).
- Northern Territory: the core vegetation attributes of species, strata, height and cover, and the methods used in Brocklehurst *et al.* (2007) and Russell-Smith (1991).
- Queensland: the structural data attributes (cover and height) from Webb (1978). Floristic information is added if possible, using the core vegetation attributes of strata, height, basal area and cover described earlier in this chapter. A field key to structural types of Australian rainforest vegetation can be found in Neldner *et al.* (2022), from Webb (1978).
- Western Australia: the core vegetation attributes of species, strata, height and cover, and the methods used previously outlined in McKenzie (1991), Corey *et al.* (2013) and Kenneally (2018).

To describe rainforests using the Webb (1978) structural types, additional specific attributes need to be collected. These are used in the key provided in Neldner *et al.* (2022). In contrast to the other core attributes outlined in this chapter, these are generally recorded as categorical scores in the broad classes of: 'absent', 'uncommon and/or inconspicuous', 'occasional or uncommon but conspicuous', 'abundant or common' (Webb *et al.* 1976; Neldner *et al.* 2022). Additional structural features that may require data collection include buttresses, leaf and stem size.

Buttresses

Buttresses are flanges at the bases of trees and there are two types recognised: plank buttresses and spur buttresses. Plank buttresses are defined as 'parallel-sided, relatively thin and roughly triangular flanges or "fins" which extend at least 1 m up the trunk' (Webb 1978). Spur buttresses are *thick and rounded* (Webb 1978). Buttresses that are <1 m long are considered spur buttresses.

Leaf and stem size

Leaf size classes for describing tropical and subtropical rainforests are based on the sizes of the leaves of the canopy (excluding emergents). Precise calculation of leaf area is not required. Visual inspection is most often all that is required to determine common leaf size (Webb 1978). However, if this is difficult, record the leaf length and width (Figure 14) of a sample of 10 canopy trees. Leaf size rules and categories, and the terms for describing leaf size in the tallest stratum of tropical and subtropical rainforests, are given in Table 23.

Two possible, but unlikely, combinations of leaf sizes cannot be described adequately by this scheme. If all leaf sizes are represented equally (20% each), the forest should be described as notophyll. If any three size classes are represented equally (for example 30% macrophyll, 30% mesophyll and 30% notophyll), the intermediate leaf size term mesophyll should be

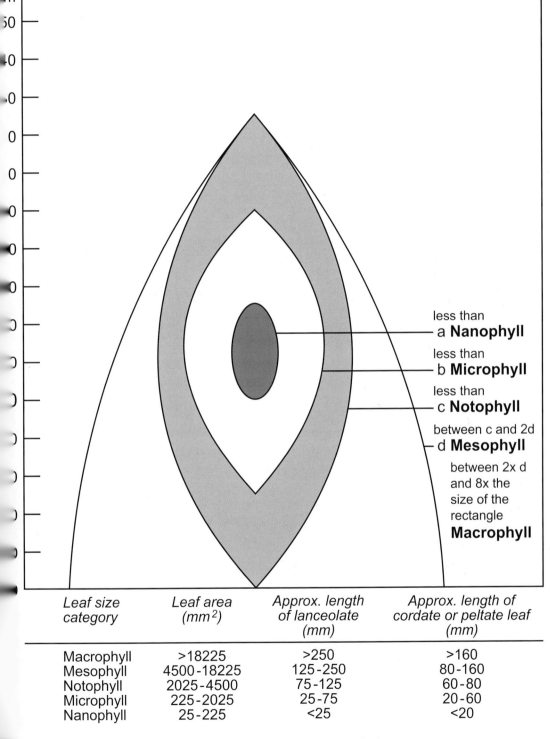

less than
a **Nanophyll**

less than
b **Microphyll**

less than
c **Notophyll**

between c and 2d
d **Mesophyll**

between 2x d
and 8x the
size of the
rectangle
Macrophyll

Leaf size category	Leaf area (mm^2)	Approx. length of lanceolate (mm)	Approx. length of cordate or peltate leaf (mm)
Macrophyll	>18225	>250	>160
Mesophyll	4500-18225	125-250	80-160
Notophyll	2025-4500	75-125	60-80
Microphyll	225-2025	25-75	20-60
Nanophyll	25-225	<25	<20

e 14 Actual leaf size categories for rainforest trees and rules to be applied.

Table 23   Leaf size rules and categories, and the terms for describing leaf size in the tallest stratum of tropical and subtropical rainforests

Term describing leaf size of forest stand	Number of individual trees (maximum 10) with specified leaf sizes	Percentage of individuals in tallest stratum with specified leaf size
Macrophyll	>5 macrophyll	>50% macrophyll
Macrophyll–mesophyll	3–5 macrophyll and 1–4 mesophyll	30–50% macrophyll and 10–40% mesophyll
Mesophyll	>5 mesophyll	>50% mesophyll
Mesophyll–notophyll	3–5 mesophyll and 1–4 notophyll	30–50% mesophyll and 10–40% notophyll
Notophyll	>5 notophyll	>50% notophyll
Notophyll–microphyll	3–5 notophyll and 1–4 microphyll	30–50% notophyll and 10–40% microphyll
Microphyll	>5 microphyll	>50% microphyll
Microphyll–nanophyll	3–5 microphyll and 1–4 nanophyll	30–50% microphyll and 10–40% nanophyll
Nanophyll	>5 nanophyll	>50% nanophyll

Adapted from Hnatiuk *et al.* (2009).

selected. To determine uniformity of stem size, a visual assessment of the canopy trees is sufficient. Stem diameters are described as 'usually uneven in size' or 'usually uniform in size'.

Rules for describing leaf size in the tallest stratum of tropical or subtropical rainforests are as follows:

- Where the average leaf size of a tree appears to be intermediate between size classes (e.g. the leaf length of a lanceolate leaf is approximately 75 mm), the larger size class should be nominated.
- Only leaves that are exposed to the sun should be considered. Because these leaves are usually at the top of a tree, a shotgun or catapult may be necessary. An alternative is to locate recently fallen leaves on the ground. The leaf sizes of shaded trees in the lower layers are frequently larger than those of the canopy species and therefore may lead to an overestimate of leaf size.
- Leaves of palms, aroids and vines should not be considered.
- The size of the leaflet of a compound leaf should be considered.

Rainforest indicator growth form

Many forests develop strata that are visually dominated by particular growth forms. Refer to illustrations in Webb *et al.* (1976) and record the relative abundance of each of these. Table 24 lists the indicator growth forms.

Table 24   Rainforest indicator growth forms

Climbing pandans	Vines – thorns/prickles/hooks	Seedlings
Climbing aroids	Multi-stem plants	Tree ferns
Epiphytes on tree trunks	Single-stem plants	Bamboo
Epiphytes on tree crowns	Fan palms	Ground aroids
Mosses (replacing epiphytes in canopy and high up tree trunks)	Banyans	Ground ferns
Robust lianes (>5 cm stem diameter)	Stranglers	Mosses (on ground)
Slender lianes (<5 cm stem diameter)	Pandans	Lichens
Vines – feather palm leaves	Shrubs	Feather palms

Rainforest leaf-fall characteristic

This is the degree of deciduousness of a patch of rainforest vegetation determined by the leaf-fall characteristics of the canopy species. The leaf-fall class is determined as the proportion (recorded as a percentage) of each of deciduous, semi-deciduous, semi-evergreen and strictly evergreen species in the canopy. These leaf-fall classes are defined in Webb (1978) as follows:

- 'Deciduous' means species or certain individuals of a species that obligatorily lose their leaves completely each year.
- 'Semi-deciduous' indicates that most leafless species are truly deciduous but that some are facultative (i.e. leaf fall is controlled by the severity of the dry season).
- 'Semi-evergreen' means that few or none of the species are truly deciduous and that most of those that shed their leaves do so incompletely depending on the severity of the dry season.
- 'Evergreen' means species that do not lose leaves in a seasonal pattern.

Oceanic cool temperate rainforests

Oceanic cool temperate rainforests are known to occur in Victoria and Tasmania (Keith *et al.* 2020). There are several methods that can be employed to collect, classify and describe these vegetation types. For Victoria's temperate rainforests, the method employed to characterise vegetation types follows the same core field attributes as for non-rainforest native vegetation types. For Tasmania's oceanic cool temperate rainforests, field attributes include a combination of floristics and structure, akin to the core attributes detailed in this chapter. However, Tasmania has established a classification system that is only employed in that state (Jarman *et al.* 1984, 1999). This system uses a combination of floristics and structure that can be coded into the NVIS vegetation hierarchy at level 5 (see Table 14). The key to these communities is available in relevant regional modules of the online *Forest Botany Manual* (Forest Practices Authority 2005).

To ensure sampled data for Tasmanian oceanic cool temperate rainforests can be assigned to a described community, the standard vegetation survey methods described in this chapter should be followed. They must include a record of plant species and abundance, together with the core structural attributes of each stratum.

For further information, refer to the texts herein, and to determine equivalence units between the Tasmanian Regional Forest Agreement (RFA) and TASVEG (the state-wide vegetation map), refer to the TASVEG online companion manual (Kitchener and Harris 2013).

WETLANDS

Wetlands and aquatic ecosystems are important and recognisable features of Australian landscapes. They support a range of biotic and abiotic processes that contribute significantly to the function of land and seascapes and the maintenance of ecological processes and biodiversity (DES 2021). The definition and classification of wetlands is complex and in many instances is linked to a specific purpose or application (Cowardin *et al.* 1979). They encompass a wide range of ecosystems and habitats, some of which are not immediately associated with 'wetlands' in the broad sense. At their upper limits they represent the transition between aquatic and terrestrial/subterranean ecosystems, and at their lower limit (at least in part), the transition from the neritic zone to the oceanic zone. The characterisation of wetlands requires the use of the Field Handbook, as wetlands are defined by a combination of soil, landscape and vegetation features (see below).

Background and definitions

In the previous edition of the Field Handbook, Hnatiuk *et al.* (2009) included a discussion of wetlands and a recommended typology of Australian aquatic and wetland types. The discussion included a definition of wetlands, in line with the RAMSAR convention (Anon. 1994), and recommended typology for Australian wetlands as outlined in *A Directory of Important Wetlands in Australia* (Commonwealth of Australia 2001). However, as indicated by the names and descriptions of these types, the identification of a wetland depends on attributes other than vegetation, thus vegetation may not be an important (or even present) feature of a type, as defined in this typology.

In this edition, the concept of wetlands is expanded to include 'aquatic ecosystems' and is defined in alignment with the *Interim Australian National Aquatic Ecosystem Toolkit* (Aquatic Ecosystems Task Group 2012b). Wetlands (including aquatic ecosystems) are defined as ecosystems dependent on flows, or periodic or sustained inundation/waterlogging (including fresh, brackish or salt water), for their ecological integrity (e.g. wetlands, rivers, karst and other groundwater-dependent ecosystems, saltmarshes, estuaries, and areas of marine water the depth of which at low tide does not exceed 6 metres).

To be defined as a wetland or aquatic ecosystem, the area must typically have one or more of the following attributes (Department of Environment and Resource Management 2011):

- The area is typically subject to permanent, temporary inundation, or saturation.
- The land supports plants or animals that are adapted to and dependent on living in wet conditions for at least part of their life cycle.
- The substratum is predominantly undrained soils that are saturated, flooded or ponded long enough to develop anaerobic conditions in the upper layers.
- The substratum is not soil and is saturated with water or covered by water at some time.

Framework for wetland description

There are numerous methods for the classification of wetlands and aquatic ecosystems which may place a differential emphasis on the importance of vegetation in the classification system (e.g. WetlandInfo). As with other forms of classification (vegetation, soil), it must be preceded by adequate and standardised description.

Although recognising the importance of vegetation in contributing to the classification of wetlands and aquatic ecosystems, 'wetland vegetation' is a function of the climate and hydrology of a system. Consequently, the classification of wetlands relies upon numerous characteristics of the biotic and abiotic environment and is not solely linked to vegetation.

In this edition, we adopt a similar wetland classification as proposed by the Australian Aquatic Ecosystems Task Group (AQETG) in the Aquatic Ecosystems Toolkit and the Australian National Aquatic Ecosystem (ANAE) classification framework (Aquatic Ecosystems Task Group 2012a and b).

The ANAE classification framework provides a hierarchy for the attribution of wetlands and aquatic ecosystems that draws on a range of data sources at various scales, from the regional to the 'site', or in some instances 'sub-site' (habitat). The depth of the hierarchy is largely dependent upon the resolution and purpose of the classification schema being applied.

Broad scale attributes (Level 1 (regional) and 2 (landscape)) typically utilise existing broad scale, high-level regionalisations to characterise aquatic ecosystems at the national and/or regional level. These levels of the classification are driven by both aquatic and terrestrial components of the landscape, such as landform, climate, hydrology, topography and water influence. Level 1 and to some degree Level 2, provide broad context based on collated, existing datasets and inferred patterns across a variety of spatial scales, typically at the jurisdictional or national scale. These levels of classification are essential in providing context for any assessment of wetlands and aquatic ecosystems and are largely not reliant upon data collected in the field.

Level 2 builds on Level 1 attributes by providing a finer-scale aquatic ecosystem regionalisation. This is based on attributes that are relevant at a landscape scale providing increased contextual information for an ecosystem. Level 3 (Class, System and Habitat) attributes focus on those aspects of the landscape that are dependent on water and are most likely to be relevant at the 'site' or 'sub-site' scale of data collection and measurement. Aquatic ecosystems are separated into two major categories ('class'): surface waters and subterranean. These are then further broken down into major aquatic 'systems' based on those defined by Cowardin *et al.* (1979) for surface waters (marine, estuarine, lacustrine, palustrine, riverine and floodplain), and Tomlinson and Boulton (2010) for subterranean systems (fractured, porous sedimentary rock, unconsolidated and cave/karst).

Below the level of 'system', a pool of attributes based on ecological theory and their general use for aquatic ecosystem mapping have been proposed by AQETG to characterise each system. Collectively, these attributes identify the 'habitats'. Section 3.4 of the Interim ANAE classification framework (Aquatic Ecosystems Task Group 2012b) provides detailed definitions for the wetland and aquatic 'systems'. Within these systems, site level attributes collected in the field can be used to characterise 'habitats'. This is a finer level of detail that may be relevant to the practical characterisation and management of the wetland and aquatic ecosystem.

Elements of wetland characterisation

Measures collected at the site scale are integral to the classification of wetlands and aquatic ecosystems and are relevant at a number of levels in the ANAE classification framework. However, they are most relevant at Level 3 in the hierarchy and more specifically in the assignment of 'systems' and the characterisation of 'habitats'. Levels 1 and 2 of the classification should identify at the broadest scale the distinction between the 'classes' and 'systems' identified at Level 3 in the hierarchy. Site scale information will allow refinement of the concepts associated with 'systems' and further classification of these ecosystems to the 'habitat' level.

Field measures detailed in this Field Handbook, across a range of disciplines, are of particular relevance to the classification of wetlands and aquatic ecosystems under the ANAE classification framework (Table 25).

NON-NATIVE VEGETATION

As for native vegetative types, the same core field attributes are relevant for classifying and describing non-native vegetation types. Rather than reiterating the core attributes required,

Table 25 Suggested field attributes required to describe and classify wetlands, including ground water dependent ecosystems, contained in this Field Handbook, and from other sources

Table 1 in Module 2 of the ANAE (Aquatic Ecosystems Task Group 2012b) provides further information and references relevant to each attribute

System	Wetland and aquatic ecosystem attribute	Relevant Field Handbook attributes	Other	Notes
Estuarine	Substrate	Substrate All attributes Land surface Coarse fragments	Benthic biophysical variables	Geoscience Australia (https://www.ga.gov.au/scientific-topics/marine/seabed-mapping) (https://portal.ga.gov.au/persona/marine)
	Structural Macrobiota	Vegetation Growth form Structure Floristics	Macroalgae Coral Filter feeders	Geoscience Australia (https://www.ga.gov.au/scientific-topics/marine/seabed-mapping) (https://portal.ga.gov.au/persona/marine)
	Water depth	Landform* Pattern Element Land surface* Elevation Drainage height Depth to free water Runoff	Bathymetric data	Geoscience Australia (https://www.ga.gov.au/scientific-topics/marine/seabed-mapping) (https://portal.ga.gov.au/persona/marine)
	Light availability	NA		
	Nutrient availability	NA		
	Exposure		Seabed energy	Geoscience Australia (https://www.ga.gov.au/scientific-topics/marine/seabed-mapping) (https://portal.ga.gov.au/persona/marine)
	Water type	NA	Salinity	
	Water influence	NA	Tide/wave/river	Geoscience Australia (https://portal.ga.gov.au/persona/marine) OzCoasts (https://ozcoasts.org.au/)
	Enclosure	Landform Pattern Element		

System	Wetland and aquatic ecosystem attribute	Relevant Field Handbook attributes	Other	Notes
Palustrine/ Lacustrine/Riverine/ Floodplain	Landform	Landform Pattern Element Land surface Slope Elevation Drainage height Disturbance Microrelief Depth to free water Runoff		
	Confinement**	Landform Pattern Element Substrate Properties of substrate material Properties of substrate masses		
	Soil	Soil profile All attributes		
	Substrate	Substrate All attributes Land surface Coarse fragments		
	Vegetation	Growth form Structure Height Cover		
	Water source	Landform Pattern? Element? Land surface Slope Runoff		May be some inference from landform pattern and element as to potential water sources.

Table 25 (cont.)

System	Wetland and aquatic ecosystem attribute	Relevant Field Handbook attributes	Other	Notes
	Water type	NA	Salinity pH	
	Water regime	NA Land surface Inundation Soil profile	Presence of water Duration of water Seasonality of water	Digital Earth Australia (https://www.dea.ga.gov.au/products/dea-water-observations) May be some inference from soil profile as to frequency and duration of any saturation within the profile.
Subterranean	Confinement	Land surface (Inundation) Landform Substrate		
	Dominant porosity	Substrate (properties of substrate material)		
	Water type	NA	Salinity pH Stratification	
	Residence time	NA		
	Dominant recharge sources	NA		
	Saturation regime	Soil profile Substrate Vegetation		
	Hydraulic conductivity	NA		
	Groundwater to surface water connectivity regime	NA		

*For tidal depth classes only.
**Riverine only.

Table 26 National non-native vegetation classification systems and relevant guidelines and field manuals

Classification system	Citation	URL
National		
NVIS Ecological/Land Cover Hierarchy	NVIS Technical Working Group (2017)	https://www.awe.gov.au/agriculture-land/land/publications/australian-vegetation-attribute-manual-version-7
National Land Use and Cover Mapping (ACLUMP) ALUM Classification	ABARES (2016)	https://www.awe.gov.au/abares/aclump/about-aclump https://www.awe.gov.au/abares/aclump/land-use/alum-classification/alum-classes
Additional manuals		
	Abed and Stephens (2003)	https://www.agriculture.gov.au/abares/forestsaustralia/publications/tree-measurement-manual
Farm Forestry	DAWE (2022)	https://www.awe.gov.au/agriculture-land/forestry/australias-forests/plantation-farm-forestry https://www.awe.gov.au/sites/default/files/documents/farm-forestry-growing-together.pdf
Field measurement of fractional ground cover	Muir *et al.* (2011)	
Weeds of National Significance		https://weeds.org.au/wp-content/uploads/2020/04/Weeds_Manual.pdf

this section provides examples of national classification systems for non-native vegetation, including land cover and/or land use mapping. The purpose of this section is not to recommend a classification system to use in the context of this chapter, instead it identifies the current and operational classification systems and methods that are employed nationally. Table 26 provides examples of non-native vegetation and land cover classification systems, and each is summarised in this section. Also included in the table are references to other guidelines and field manuals where additional field metrics may be required (i.e. farm forestry and surveying weeds of national significance).

DESCRIPTION OF VEGETATION USING THE NVIS FRAMEWORK

The core vegetation attributes collected following the standards outlined in this chapter can be used to describe and classify the vegetation type at a site in a consistent manner. While these descriptions may differ between jurisdictions, they are derived following the same process. Here, the NVIS terminology (NVIS Technical Working Group 2017) is used as an example for characterising the vegetation type at a site. This terminology was adapted from Specht (1970), Specht *et al.* (1974) and Walker and Hopkins (1990).

The initial step is to determine the vegetation *structural formation*, which is a standardised terminology used to integrate *growth form*, *height* and *cover* within each *stratum/substratum*. The allocation of a height class to a growth form in a stratum/substratum gives rise to a particular height qualifier, as provided in Table 20. The height qualifier and growth form are then added to a cover class using the applicable cover type to define the structural formation, provided in Table 27. The structural formation is used in generating levels 1 to 5 of the NVIS

Table 27 NVIS cover characteristics, growth forms, height ranges and corresponding structural formation classes

Cover characteristics

	70–100	30–70	10–30	<10	≈0	0–5	unknown
Foliage cover %	70–100	30–70	10–30	<10	≈0	0–5	unknown
Crown cover %	>80	50–80	20–50	0.25–20	<0.25	0–5	unknown
Cover %	>80	50–80	20–50	0.25–20	<0.25	0–5	unknown
Cover code	d	c	i	r	bi	bc	

Structural formation classes

Growth form	Height ranges (m)							
tree, palm	<10, 10–30, >30	closed forest	open forest	woodland	open woodland	isolated trees	isolated clumps of trees	trees
tree mallee	<3, <10, 10–30	closed mallee forest	open mallee forest	mallee woodland	open mallee woodland	isolated mallee trees	isolated clumps of mallee trees	mallee trees
shrub, cycad, grass tree, tree fern	<1, 1–2, >2	closed shrubland	shrubland	open shrubland	sparse shrubland	isolated shrubs	isolated clumps of shrubs	shrubs
mallee shrub	<3, <10, 10–30	closed mallee shrubland	mallee shrubland	open mallee shrubland	sparse mallee shrubland	isolated mallee shrubs	isolated clumps of mallee shrubs	mallee shrubs
heath shrub	<1, 1–2, >2	closed heathland	heathland	open heathland	sparse heathland	isolated heath shrubs	isolated clumps of heath shrubs	heath shrubs
chenopod shrub	<1, 1–2, >2	closed chenopod shrubland	chenopod shrubland	open chenopod shrubland	sparse chenopod shrubland	isolated chenopod shrubs	isolated clumps of chenopod shrubs	chenopod shrubs
samphire shrub	<0.5, >0.5	closed samphire shrubland	samphire shrubland	open samphire shrubland	sparse samphire shrubland	isolated samphire shrubs	isolated clumps of samphire shrubs	samphire shrubs
hummock grass	<2, >2	closed hummock grassland	hummock grassland	open hummock grassland	sparse hummock grassland	isolated hummock grasses	isolated clumps of hummock grasses	hummock grasses
tussock grass	<0.5, >0.5	closed tussock grassland	tussock grassland	open tussock grassland	sparse tussock grassland	isolated tussock grasses	isolated clumps of tussock grasses	tussock grasses

		closed grassland	grassland	open grassland	sparse grassland	isolated grasses	isolated clumps of grasses	other grasses
other grass	<0.5, >0.5	closed grassland	grassland	open grassland	sparse grassland	isolated grasses	isolated clumps of grasses	other grasses
edge	<0.5, >0.5	closed sedgeland	sedgeland	open sedgeland	sparse sedgeland	isolated sedges	isolated clumps of sedges	sedges
rush	<0.5, >0.5	closed rushland	rushland	open rushland	sparse rushland	isolated rushes	isolated clumps of rushes	rushes
forb	<0.5, >0.5	closed forbland	forbland	open forbland	sparse forbland	isolated forbs	isolated clumps of forbs	forbs
fern	<1, 1–2, >2	closed fernland	fernland	open fernland	sparse fernland	isolated ferns	isolated clumps of ferns	ferns
bryophyte	<0.5	closed bryophyteland	bryophyteland	open bryophyteland	sparse bryophyteland	isolated bryophytes	isolated clumps of bryophytes	bryophytes
lichen	<0.5	closed lichenland	lichenland	open lichenland	sparse lichenland	isolated lichens	isolated clumps of lichens	lichens
vine	<10, 10–30, >30	closed vineland	vineland	open vineland	sparse vineland	isolated vines	isolated clumps of vines	vines
aquatic	0–0.5, <1	closed aquatic bed	aquatic bed	open aquatic bed	sparse aquatics	isolated aquatics	isolated clumps of aquatics	aquatics
seagrass	0–0.5, <1	closed seagrass bed	seagrassbed	open seagrassbed	sparse seagrassbed	isolated seagrasses	isolated clumps of seagrasses	seagrasses

Source: NVIS Technical Working Group (2017).

Table 28 Specifications for the NVIS vegetation hierarchy to generate a vegetation type description using a tropical savanna example

Level	Description	Combined requirements	Genera and/or species	Growth form	Cover	Height
1	Class	Dominant growth form for the structurally dominant stratum. Example: **Tree**	–	One dominant growth form for the structurally dominant stratum	–	–
2	Structural formation	Dominant growth form, cover and height for the structurally dominant stratum. Example: **Low Open Woodland**	–	One dominant growth form for the dominant stratum	One cover class for the dominant stratum	One height class for the dominant stratum
3	Broad floristic formation	Dominant genus (or genera) plus growth form, cover and height for the structurally dominant stratum. Example: **Corymbia Low Open Woodland**	One or two dominant genera for the dominant stratum or one genus with the word '(mixed)' for the dominant stratum	One dominant growth form for the dominant stratum	One cover class for the dominant stratum	One height class for the dominant stratum
4	Sub-formation	Dominant genus (or genera) plus growth form, cover and height for each of the three main strata (i.e. U, M and G). Example: **U + Corymbia Low Open Woodland / Acacia Mid Sparse Shrubland / Triodia Open Hummock Grassland**	One or two dominant genera for each stratum (max three strata; i.e. for U, M, G where substantially present)	One dominant growth form for each stratum (max three strata; i.e. for U, M, G where present)	One cover class for each stratum (max three strata)	One height class for each stratum (max three strata)
5	Association	List up to three growth forms and three species in decreasing order of dominance for the three major strata (i.e. U, M and G). Each stratum has a structural formation. Dominant stratum indicated with '+'. Species listed in order of dominance. Example: **U + Corymbia dichromophloia, Corymbia ferruginea, Erythrophleum chlorostachys Low Open Woodland / M Acacia monticola, Grevillea refracta, Acacia difficilis Mid Sparse Shrubland / G Triodia pungens, Schizachyrium fragile, Acacia spondylophylla Mid Open Hummock Grassland**	Up to three species for each stratum (max three strata; i.e. for U, M, G where present). Indicate dominant genus, genera or mixed.	Up to three growth forms for each stratum (max three strata; i.e. for U, M, G where present)	One cover class for each stratum (max three strata; i.e. for U, M, G where present)	One height class for each stratum (max three strata; i.e. for U, M, G where present)

Level	Description	Combined requirements	Genera and/or species	Growth form	Cover	Height
6	Sub-association	List up to five growth forms and five species in decreasing order of dominance for up to nine substrata. Each stratum has a structural formation. Dominant stratum indicated with '+'. Species listed in order of dominance. Example: **U1 + Corymbia dichromophloia, Corymbia ferruginea, Erythrophleum chlorostachys Low Open Woodland / M1 Acacia monticola, Grevillea refracta, Acacia difficilis, Grevillea wickhamii, Acacia stipuligera Mid Sparse Shrubland / G1 Triodia pungens, Schizachyrium fragile, Acacia spondylophylla, Aristida holathera, Eriachne ciliata Mid Open Hummock Grassland**	Up to five species for each substratum (i.e. for U1, U2, U3, M1, M2, M3, G1, G2, G3 where present). Indicate dominant genus, genera or mixed.	Up to five growth forms for each substratum	One cover class for each substratum	One height class for each substratum

Adapted from NVIS Technical Working Group (2017).

Level 1, Class = tree   **Level 2**, Structural formation = low open woodland

Level 3, Broad floristic formation = *Corymbia* low open woodland

Level 4, Sub-formation = +*Corymbia* low open woodland / *Acacia* mid-sparse shrubland / *Triodia* mid-open hummock grassland

Association (Level 5) and Sub-association (Level 6).

Level 5, Association = U+*Corymbia dichromophloia, Corymbia ferruginea, Erythrophleum chlorostachys* low open woodland / M *Acacia monticola, Grevillea refracta, Acacia difficilis* mid-sparse shrubland / G *Triodia pungens, Schizachyrium fragile, Acacia spondylophylla* mid-open hummock grassland

Level 6, Sub-association = U1+*Corymbia dichromophloia, Corymbia ferruginea, Erythrophleum chlorostachys* low open woodland / M1 *Acacia monticola, Grevillea refracta, Acacia difficilis, Grevillea wickhamii, Acacia stipuligera* mid-sparse shrubland / G1 *Triodia pungens, Schizachyrium fragile, Acacia spondylophylla, Aristida holathera, Eriachne ciliata* mid-open hummock grassland

Figure 15   Example low open woodland from the tropical savanna biome in northern Australia.

vegetation hierarchy (see Table 14), and thus become a relatively user-friendly summary of the dominant *growth form*, *cover* and *height* for a detailed vegetation type description at level 6 (sub-association). A dominant genus is added at level 3 (broad floristic formation) for the dominant stratum, then again at level 4 (sub-formation) for all strata recognised at a site. Levels 5 (association) and 6 add floristics, up to three dominant species per stratum/substratum for level 5 and up to five dominant species for level 6.

Table 28 details the specifications (including examples) to put a site-based NVIS vegetation type description together at the six levels of the NVIS vegetation hierarchy. The NVIS classification framework uses codes for growth form, cover class, height class, dominance classifiers etc. However, in this case, a plain language description is more user friendly and interpretable, thus rather than using codes, the full structural formation (Table 27) is inserted into the vegetation type description. Strata and substrata are separated by a forward slash (/), the dominant stratum is indicated with a plus sign (+), and species dominance is listed in order of dominance ranked by the cover per cent values. Table 28 provides an example of a woodland from the tropical savanna biome in northern Australia classified to six levels of the NVIS vegetation hierarchy (Figure 15).

VEGETATION CONDITION

Condition, by definition, is a value-based consideration, rather than an inherent property of vegetation. So, before we can decide what measures may be relevant and how we capture or combine them, we first need to articulate the values we are attempting to represent (i.e. condition for what?). In Australia, the values and associated approaches to assessing vegetation condition usually fall into two categories: utilitarian values and intrinsic (conservation) values (Keith and Gorrod 2006).

Utilitarian values
Utilitarian values are where condition of vegetation is based on the provision of services to people (e.g. carbon sequestration, water quality and fodder provision) and associated accounting schemes (e.g. System of Environmental Economic Accounting).

Intrinsic values
Intrinsic (conservation) values are benchmarked in very different ways, including:

> *Generic 'naturalness'* approaches, where specific conservation values are not considered, and the same indicators are measured for each vegetation type but are evaluated against different benchmarks (using historic or best-available examples). An inherent assumption in such assessments is that deviations from a benchmark are always undesirable, which presents challenges when vegetation goes through typical cycles of disturbance dynamics. Many existing state condition assessment schemes use this approach, combining measures based on various structural and habitat features, species or lifeform diversity, weediness, and regeneration (sometimes in combination with conservation status and generic landscape metrics). Typical examples include Habitat Hectares in Victoria and Bushland Condition Monitoring in South Australia. The national Habitat Condition Assessment System (HCAS) is a related approach that has the same underlying premise but combines site measures with remote sensing.

Table 29   Vegetation condition manuals for Australian jurisdictions, including national, state and territory

Jurisdiction	Name	Citation	URL
National	CSIRO - HCAS	Williams *et al.* (2021)	https://publications.csiro.au/publications/publication/PIcsiro:EP2021-1200
		Harwood *et al.* (2016)	https://research.csiro.au/biodiversity-knowledge/projects/hcas/
		Donohue *et al.* (2014)	https://publications.csiro.au/publications/publication/PIcsiro:EP1311716
Australian Capital Territory	Adapted BioNet ACT Vegetation Types ACT Vegetation Map	Capital Ecology (2018) ACT Government (2015)	https://www.planning.act.gov.au/__data/assets/pdf_file/0005/1436369/2017-woodland-quality-and-extent-mapping-act-government-environmental-offsets.pdf https://www.environment.act.gov.au/__data/assets/pdf_file/0005/728600/Schedule-1-Environmental-offsets-Assessment-Methodology-FINAL-2.pdf#:~:text=The%20ACT%20Environmental%20Offsets%20Assessment%20Methodology%20%28the%20methodology%29,it%20is%20most%20cost%20effective%20to%20do%20so
New South Wales	BioNet		https://www.environment.nsw.gov.au/topics/animals-and-plants/native-vegetation/vegetation-condition-benchmarks
Northern Territory	Condition Assessment	Price and Brocklehurst (2008)	https://hdl.handle.net/10070/673751
Queensland	BioCondition	Eyre *et al.* (2017)	https://www.qld.gov.au/__data/assets/pdf_file/0027/68571/reference-sites-biocondition.pdf
South Australia	Condition Indicators	Native Vegetation Council (2020a,b,c)	https://cdn.environment.sa.gov.au/environment/docs/scattered_tree_assessment_manual_1_july_2020.pdf https://cdn.environment.sa.gov.au/environment/docs/bushland_assessment_manual_1_july_2020.pdf https://cdn.environment.sa.gov.au/environment/docs/rangelands_assessment_manual_1_july_2020.pdf
Tasmania		Michaels *et al.* (2020)	https://nre.tas.gov.au/conservation/development-planning-conservation-assessment/planning-tools/monitoring-and-mapping-tasmanias-vegetation-(tasveg)/vegetation-condition-monitoring
Victoria	Habitat Hectares	DSE (2004)	https://www.environment.vic.gov.au/__data/assets/pdf_file/0016/91150/Vegetation-Quality-Assessment-Manual-Version-1.3.pdf
		Parkes *et al.* (2003)	https://urldefense.com/v3/__https:/onlinelibrary.wiley.com/doi/full/10.1046/j.1442-8903.4.s.4.x__;!!PUY2jUP3Fp7oEg!G3zHYOMtvIRgey1NQfY7t7buYhKtWa5VxBFLSOo8Q96nZd4p9LRo8cfJdlTFtSNf4Za0lFqrAn-rjEAv428YBV91nCtpZ1f51nonDvTdug$
Western Australia	Condition Assessment	Casson *et al.* (2009)	https://library.dbca.wa.gov.au/static/FullTextFiles/926887.pdf

Specific 'outcome' approaches, where the indictors that are measured vary, but are derived from a model and/or evidence that directly relates them to the specific values of conservation concern. Holistic system-level examples include the Australian Ecosystems Models Framework or other state-transition approaches (such as those applied to Box Gum Woodlands in Eastern Australia). There are also many examples that are focussed on the provision of habitat for species of conservation concern, which may even encompass areas that are highly modified (from a 'naturalness' perspective), but nevertheless support values of high conservation significance. Many individual threatened species recovery programs provide details of habitats and various condition measures. There are also an increasing number of multi-species condition assessments (such as the Dynamic Fire and Biodiversity Tool in South Australia). These outcome-based approaches usually consider disturbance dynamics, enabling changes in specific vegetation measures to be directly linked to the values of concern, supporting clear interpretations.

There have been numerous methodologies to assess vegetation condition in Australia, at both national and state levels. This section does not attempt to provide one standard for assessing vegetation condition based on field attributes. Rather, it provides the references to the various Australian state and territory methods to assess vegetation condition (Table 29). The core attributes outlined in this chapter can be used as the metrics for assessing change in structure and floristics to assess against benchmark criteria. Table 29 also includes national scale initiatives, such as HCAS for Australia which uses remote sensing, spatial ecology modelling and data from on-ground condition assessments to generate a national view of condition.

LAND SURFACE

RC McDonald, RF Isbell and JG Speight

This section is concerned mainly with surface phenomena affecting land use and soil development that have traditionally been noted at the point of soil observation. Most of the attributes described for the land surface have implications regarding the use of land; some may also reflect significant processes occurring within the soil (e.g. microrelief) or landscape. Some attributes (e.g. disturbance, erosion) may reflect the influence of present or past land use practice, but it is important that their status be recorded at a known point in time.

It may be difficult to estimate accurately some other required attributes (e.g. inundation), but the field observer usually has the benefit of some local experience and is better placed to make such an estimate than a subsequent user of the data.

ASPECT
Give aspect as compass bearing to nearest 10 degrees. On level lands (less than 1% slope), aspect need not be recorded.

ELEVATION
Method of determination of elevation

L   *Levelled from survey datum or estimated from contour plan (1:10 000 or larger scale)*

M   *Interpolated from contour map with contour interval of 20 m or less*

A   *Determined by altimeter*

E   *Estimate*

R   *Determined from remote sensing*

Elevation value
Give elevation in metres above sea level.

DRAINAGE HEIGHT
This is the height of the point of soil observation above the flat, depression or stream bed that forms the effective bottom of the toposequence (page 20).

Means of evaluation of drainage height

As in Elevation above.

Drainage height value

Give drainage height in metres.

DISTURBANCE OF SITE

The disturbance of a site is important contextual information that has implications for the use and interpretation of site data. Broad categories of disturbance are defined, which users may subdivide where necessary. Land use (which is different to disturbance) may be recorded using codes associated with the Australian Land Use and Management Classification (ABARES 2016) or the related National Land Management Practices Classification System (under development).

0	*No effective disturbance; natural*
1	*No effective disturbance other than grazing by hoofed animals*
2	*Limited clearing (e.g. selective logging, <50% of trees cleared)*
3	*Extensive clearing (e.g. poisoning, ringbarking)*
4	*Complete clearing; pasture, native or improved, but never cultivated*
5	*Complete clearing; pasture, native or improved, cultivated at some stage*
6	*Cultivation; rainfed*
7	*Cultivation; irrigated, past or present*
8	*Highly disturbed (e.g. quarrying, road works, mining, landfill, urban)*
9	*Regrowth after clearing*
10	*Significant natural disturbance (e.g. cyclonic impact, fire, hail, flood)*

Method of disturbance

The method of disturbance can be recorded when relevant, most commonly in relation to mechanical disturbance, fire or flood. One or more disturbance methods may be recorded. The depth of disturbance and nature of additions to the profile may also be recorded when appropriate.

BP	*Blade ploughing*
CA	*Clay addition*
DE	*Delving*
DR	*Deep ripping*
FI	*Fire*
FL	*Flood*
LE	*Levelling*
MD	*Mole drains*

MO	*Mounding*
PI	*Profile inversion by mechanical means*
SL	*Slotting*
SP	*Spading*
TA	*Topsoil additions (e.g. turf farming)*
TO	*Topsoil stripping/grading*
WE	*Weather events (e.g. storms, cyclones)*

Depth of disturbance

X	*Unable to be determined*	
1	*Very shallow*	*<0.1 m*
2	*Shallow*	*0.1–0.3 m*
3	*Moderately deep*	*0.3–0.5 m*
4	*Deep*	*>0.5 m*

Additions to the profile, associated with disturbance

C	*Clay*
D	*Dolomite*
F	*Fertiliser*
H	*Heterogenous or variable*
L	*Lime*
M	*Manure*
O	*Organic matter*
P	*Pesticide*
S	*Sand/loam*
Y	*Gypsum*
U	*Unknown*

MICRORELIEF

Microrelief refers to relief up to a few metres about the plane of the land surface, with a horizontal dimension smaller than that of a landform element. It includes a variety of forms created through a range of mechanisms – biological and non-biological. For any given type of microrelief, attributes to be recorded include the agent of formation, dimensions, component of point observation (e.g. soil profile) and, where relevant, proportion of components.

Given the dimension of microrelief varies from sub-metre to tens of metres, the dimension of assessment of microrelief is necessarily variable. It typically would be within a 20 m radius and requires consideration of the average across the representative dimension. Microrelief density is not normally recorded but may be appropriate in some studies.

Type of microrelief

Give the type of microrelief within the site:

Z   *Zero* or *no microrelief*

Gilgai microrelief types

Gilgai is surface microrelief associated with soils containing shrink–swell clays. It does not include microrelief that results from repeated freezing and thawing, solifluction or faunal activity. Gilgai consist of mounds and/or depressions showing varying degrees of order, sometimes separated by a sub-planar or slightly undulating surface (the shelf). Further information on types of gilgai is available in Hallsworth *et al.* (1955), Beckmann *et al.* (1970) and Paton (1974).

In order of increasing dimensions, gilgai types are:

C   *Crabhole gilgai*   Irregularly distributed, small depressions and mounds separated by a more or less continuous shelf. Vertical interval usually less than 0.3 m. Horizontal interval usually 3–20 m, surface almost level.

N   *Normal gilgai*   Irregularly distributed, small mounds and subcircular depressions varying in size and spacing. Vertical interval usually less than 0.3 m, horizontal interval usually 3–10 m, surface almost level.

L   *Linear gilgai*   Long, narrow, parallel, elongate mounds and broader elongate depressions more or less at right angles to the contour (i.e. typically oriented up/down the slope). Linear to curvilinear. Usually in sloping lands. Vertical interval usually less than 0.3 m, horizontal interval usually 5–8 m.

A   *Lattice gilgai*   Discontinuous, elongate mounds and/or elongate depressions more or less at right angles to the contour. Usually in sloping lands, commonly between linear gilgai on lower slopes and plains.

M   *Melonhole gilgai*   Irregularly distributed, large depressions, usually greater than 3 m in diameter or greatest dimension, subcircular or irregular and varying from closely spaced in a network of elongate mounds to isolated depressions set in an undulating shelf with occasional small mounds. Some depressions may also contain sinkholes. The vertical interval is usually greater than 0.3 m; horizontal interval usually 5–40 m; surface almost level.

G   *Contour gilgai*   Long, elongate depressions and parallel adjacent downslope mounds, which follow the contour. These depression–mound associations are separated from each other by shelves 10–100 m wide. Depressions are up to 0.5 m deep and 30–50 m wide. Mounds are low, usually less than 0.5 m high, and often poorly defined (after Lawrie 1978).

Biotic microrelief types

This relief is caused by biotic agents, such as termite mounds, rabbit warrens, wombat burrows, pig wallows, human-made terraces, stump holes, vegetation mounds and people. For example, vegetation mounds are mounds of soil material found at the base of plants such as Dillon bush (*Nitraria billardieri*) or spinifex (*Triodia* species). Tree-tip (windthrow) mounds are another type of vegetation related microrelief. Users may subdivide types of biotic microrelief where considered necessary, such as specifying the particular species of vegetation involved in the case of vegetation mounds. Two specific types of biotic microrelief are described (these were formally listed under 'Hummocky' microrelief). Other forms of biotic microrelief are described via the agent and other relevant attributes.

D	*Debil-debil*	Small hummocks rising above a planar surface. They vary from rounded, both planar and vertically, to flat-topped, relatively steep-sided and elongate. They are usually closely and regularly spaced, ranging from 0.06 to 0.6 m in both vertical and horizontal dimensions. They are common in northern Australia on soils with impeded internal drainage and in areas of short seasonal ponding. Associated with activity of species such as earthworms (e.g. *Diplotrema planumfluvialis*) (Dyne 1987).
W	*Swamp hummock*	Steep-sided hummocks rising above a flat surface. Hummocks are frequently occupied by trees or shrubs while the lower surface may be vegetation-free or occupied by sedges or reeds. They are subject to prolonged seasonal flooding. Relief may be exaggerated by livestock.
B	*Other biotic microrelief*	This encompasses all other forms of biotic microrelief.

Other microrelief types

This relief includes types derived from a variety of processes, and in some cases multiple processes may be acting together (e.g. terracettes are a function of livestock and gravity).

F	*Frost heave*	Mounds/depressions caused by expansion of frozen soil water (also known as patterned ground).
H	*Spring hollow*	Depression associated with water flowing from rock or soil without the agency of humans.
I	*Sinkhole*	Closed depression with vertical or funnel-shaped sides.
K	*Karst microrelief*	Small depressions (<20 m diameter) in limestone country.
P	*Spring mound*	Mound associated with water currently or historically flowing from rock or soil without the agency of humans.
R	*Terracettes*	Small terraces on slopes resulting from soil creep and/or trampling by hoofed animals.
S	*Mass movement microrelief*	Hummocky microrelief on the surface of landslides, rotational slides, earth flows, debris avalanches etc. (see 'Mass movement' section below).
T	*Contour trench*	Trenches typically 0.2 m deep and 0.6 m wide, with near vertical walls, alternating with flat-crested ridges about 1.3 m wide, which extend along the contour for several metres or tens of metres. Contour trenches are known in areas in south-eastern Australia above 350 m altitude with a high effective rainfall, where they are associated with a grassland or heathland vegetation on undulating rises (compare with McElroy 1952).

U	*Mound/depression microrelief*	Undifferentiated, irregularly distributed or isolated mounds and/or depressions set in a planar surface (and are not gilgai).
V	*Spring vent*	Point of discharge of water associated with a spring. Typically an evident hole, commonly with associated small structures created by flowing sand or mud. May or may not be associated with a spring mound.
O	*Other microrelief*	

Agent of microrelief

When relevant, record the *dominant* agent of creation of the microrelief. For some naturally formed microrelief (e.g. frost heave), the agent is implicit in the type. As the agents of formation of gilgai remain unclear, no agent is recorded for them.

A	*Ant*
B	*Bird*
C	*Crab*
D	*Dissolution processes*
E	*Wind*
F	*Buffalo*
G	*Groundwater*
H	*Reptile*
I	*Ice/cold temperature (frost heave)*
J	*Tree-tip (wind thrown trees)*
K	*Domesticated livestock (sheep, goats, cattle)*
L	*Bilbies*
M	*Human*
N	*Animal (unspecified)*
O	*Other/unknown*
P	*Pig*
R	*Rabbit*
T	*Termite*
V	*Vegetation*
W	*Wombat*
Y	*Water*

Component of microrelief observed

Give the component of the microrelief for which the observation is being made – for example, the component in which the described soil profile is located. This is particularly important for gilgai. Some examples of components are provided in Figure 16.

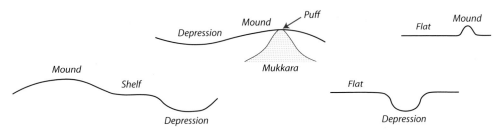

Figure 16   Examples of microrelief components. Adapted from Paton (1974).

M	*Mound*	Convex; long axis[7] not more than 3 times the shorter axis.
E	*Elongate mound*	Convex; long axis more than 3 times the shorter axis.
D	*Depression*	Concave; occurs as closed form.
L	*Elongate depression*	Concave; occurs as open-ended form.
S	*Shelf*	More or less planar surface; occurs between mounds and depressions in gilgai.
F	*Flat*	Surface in which hummocks, mounds, depressions or holes are set.
P	*Puff*	Small surface expression of subsoil material associated with mukkara.[8]
H	*Hole*	An opening in the soil surface (not a crack).
K	*Hummock*	Rises above a flat or planar surface. Sides vary from rounded to near vertical and tops vary from rounded to flat.
T	*Terrace*	Curvilinear to linear step-like feature on a sloping land surface.
O	*Other*	

Proportions of microrelief components

Give the proportions of components of microrelief within the site, thus:

A	*Equal mounds and depressions; no shelf present*
B	*More mounds than depressions; no shelf present*
C	*Fewer mounds than depressions; no shelf present*
D	*Mound, shelf and depressions all present*
E	*Depressions[9] in a planar surface; no mounds*
F	*Mounds in a planar surface; no depressions*
G	*Mounds and depressions in a planar surface*

---

7   Axis in the plane of the land surface

8   A distinct upward projection of a subsoil horizon into, and often through, the surface soil horizons. See Figure 5 in Paton (1974)

9   For codes E, F and G, the term *depressions* includes *holes* and the term *mound* includes *hummocks*

While these proportions are primarily used in the context of gilgai, they can be used for any forms of mound/depression microrelief.

Dimensions of microrelief

Record the representative dimensions of all types of microrelief within the site.

Vertical interval

Give average vertical distance, in metres, between the lowest point and a horizontal line joining the highest points of instances of the microrelief within the site.

Horizontal interval

Give average horizontal distance, in metres, between the highest points of instances of the microrelief within the site. In the case of holes (of any sort), it is the distance between the centres of holes. The depression width for holes is the diameter of the hole.

Depression width

Give horizontal distance, in metres, of the width of the lowest feature within the microrelief. The boundary of the lowest feature may be indicated by a slope inflection at its margins. Figure 17 shows the relationship between the dimensions for gilgai, but the same principles are used for other mound/depression microrelief.

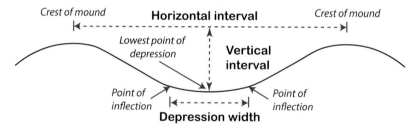

Figure 17   Dimensions relevant to gilgai microrelief.

EROSION

This section is concerned with accelerated erosion rather than natural erosion. Natural or geologic erosion is the type and rate of movement of land surface material in its undisturbed natural environment. Accelerated erosion is the more rapid erosion that follows the destruction or loss of protective cover often resulting from people's influence on the soil, vegetation or landform. It is not always easy to distinguish between accelerated and natural erosion in some landscapes where both are closely interrelated, nor is it always clear whether wind or water is the dominant agent. This is particularly so in the case of scald formation (Warren 1965).

The complexity of erosion forms creates difficulties in both defining and quantitatively estimating the extent of erosion within the dimensions of the site. Hence, the observer is advised to record the following simple parameters: state, type and severity. Assess the erosion observed at the time of the description, not the likelihood of erosion. The code **X** is used when it is probable that the erosion type is present, but it is not easily observable at the point in time (e.g. dense ground cover may be obscuring the evidence). The code **0** is used when there is complete confidence that the type of erosion is not present. *For erosion, aggradation and inundation assessment, the extent of the site is 20 m in radius.*

State of erosion

Z		No evidence of any erosion types.
A	*Active*	One or both of the following conditions apply: evidence of sediment movement; sides and/or floors of erosion form are relatively bare of vegetation.
S	*Stabilised*	One or both of the following conditions apply: no evidence of sediment movement; sides and/or floors of erosion form are revegetated.
P	*Partly stabilised*	Evidence of some active erosion and some evidence of stabilisation.

Type and severity of erosion

Record the type of erosion and the associated severity, noting that the way in which the latter is recorded varies with erosion type. For sheet, rill and gully erosion, there is no consensus in Australia on a quantitative or precise definition of what constitutes minor, moderate and severe erosion. This derives partly from the difficulty of measuring actual soil loss at a site. It also derives partly from the wide range of soils, climates and land uses, variations in any or all of which may alter the concept of minor, moderate or severe erosion.

Description of gully erosion has advanced significantly in recent times, incorporating many attributes that can be determined from remote sensing as well as via field methods (Thwaites *et al.* 2022; Thwaites *et al.* in press).

The observer wishing to record the severity of erosion may record it as minor, moderate or severe, basing the assessment on local knowledge and guided by indicators that may be present as described below (after Morse *et al.* 1987). Note the actual depth of soil loss where this can be reliably assessed.

Wind erosion W

Give presence/absence or extent of accelerated erosion.

X	*Not observed*	No wind erosion observable.
0	*None*	No wind erosion present.
1	*Minor* or *present*	Some loss of surface.
2	*Moderate*	Most or all of surface removed leaving soft or loose material.
3	*Severe*	Most or all of surface removed, leaving hard material.
4	*Very severe*	Deeper layers exposed, leaving hard material (e.g. subsoil, weathered country rock or pans).

Scald erosion C

This is the removal of surface soil by water and/or wind, often exposing a more clayey subsoil which is devoid of vegetation and relatively impermeable to water. Scalds are most common in arid or semi-arid lands.

0	*No scalding*	
1	*Minor scalding*	<5% of site scalded.
2	*Moderate scalding*	5–50% of site scalded.
3	*Severe scalding*	>50% of site scalded.

Sheet erosion S

This is the relatively uniform removal of soil from an area without the development of conspicuous channels. Indicators of sheet erosion include soil deposits in downslope sediment traps, such as fencelines or farm dams, and pedestalling, root exposure or exposure of subsoils.

X	*Not apparent*	No sheet erosion observable.
0	*No sheet erosion*	No sheet erosion present.
1	*Minor*	Indicators may include shallow soil deposits in downslope sediment traps (fencelines, farm dams). Often very difficult to assess as evidence may be lost with cultivation, pedo-turbation or revegetation.
2	*Moderate*	Indicators may include partial exposure of roots, moderate soil deposits in downslope sediment traps (fencelines, farm dams).
3	*Severe*	Indicators may include loss of surface horizons, exposure of subsoil horizons, pedestalling, root exposure, substantial soil deposits in downslope sediment traps (fencelines, farm dams).

Rill erosion R

A rill is a small channel[10] excavated by water up to 0.3 m deep, which can be largely obliterated by tillage operations (Houghton and Charman 1986).

0	*No rill erosion*	No rill erosion present.
1	*Minor*	Occasional rills.
2	*Moderate*	Common rills.
3	*Severe*	Numerous rills forming corrugated ground surface.

Gully erosion G

A gully is a channel excavated by water, more than 0.3 m deep.[11]

0	*No gully erosion*	No gully erosion present.

---

10   The word channel here does not imply a **constructed** channel for the purposes of carrying water

11   See Thwaites *et al.* (in press) for further discussion regarding the definition of a gully and metrics to define gully dimensions.

1	*Minor*	Gullies are isolated, linear, discontinuous and restricted to primary or minor drainage lines.
2	*Moderate*	Gullies are linear, continuous and restricted to primary and minor drainage lines.
3	*Severe*	Gullies are continuous or discontinuous and either tend to branch away from primary drainage lines and on to footslopes, or have multiple branches within primary drainage lines.

Gully depth

Give the maximum depth within the site using either the categories below or as a numerical value.

1	*0.3–1.5 m*
2	*1.5–3.0 m*
3	*>3 m*

Tunnel erosion T

This is the removal of subsoil by water while the surface soil remains relatively intact (Crouch 1976).

X	*Not apparent*
0	*No tunnel erosion*
1	*Present*

Stream bank erosion B

This is the removal of soil from a stream bank, typically during periods of high stream flow.

X	*Not apparent*
0	*No stream bank erosion*
1	*Present*

Wave erosion V

Erosion of beaches, beach ridges and/or dunes. This is the removal of sand or soil from the margins of beaches, beach ridges, dunes, harbours, canals, rivers, lakes or dams by wave action.

X	*Not apparent*
0	*No wave erosion*
1	*Present*

Mass movement M

A generic term for any process or sediments (mass movement deposit) resulting from the dislodgment and downslope transport of soil and rock material as a unit under direct gravitational stress. The process includes slow displacements such as creep and solifluction,[12] and rapid movements such as landslides, rock slides and falls, earthflows, debris flows and avalanches (Schoeneberger and Wysocki 2017). Agents of fluid transport (water, ice, air) may play an important, if subordinate role in the process. The terms to describe mass movement have been expanded considerably over previous editions of the Field Handbook and have been adopted primarily from Schoeneberger and Wysocki (2017). Table 30 provides the relationships between terms.

Each of the types of mass movement could be considered as landform elements or patterns, depending on their size, distribution and extent.

Types of mass movement are as follows:

Fall
Associated sediments (fall deposit) or resultant landforms (e.g. rockfall, debris fall, soil fall) characterised by very rapid movement of a mass of rock or earth that travels mostly through the air by free fall, leaping, bounding or rolling, with little or no interaction between one moving unit and another.

Flow
Associated sediments (flow deposit) and landforms characterised by slow to very rapid downslope movement of unconsolidated material that, whether saturated or comparatively dry, behaves much as a viscous fluid as it moves. Types of flows can be specified based on the dominant particle size of sediments – debris flow, earthflow (creep, mudflow), rockfall avalanche, debris avalanche.

Slide
Associated sediments (slide deposit) or resultant landforms (e.g. rotational slide, translational slide, snowslide) characterised by a failure of earth, snow or rock under shear stress along one or several surfaces that are either visible or may reasonably be inferred. The moving mass may or may not be greatly deformed, and movement may be rotational (rotational slide) or planar (translational slide). A slide can result from lateral erosion, lateral pressure, weight of overlying material, accumulation of moisture, earthquakes, expansion owing to freeze–thaw of water in cracks, regional tilting, undermining, fire and human agencies.

Spread
Associated sediments (lateral spread deposit) or resultant landforms characterised by a very rapid spread dominated by lateral movement in a soil or fractured rock mass resulting from liquefaction or plastic flow of underlying materials. Types of lateral spreads can be specified based on the dominant particle size of sediments (i.e. debris spread, earth spread, rock spread).

Topple
Associated sediments (topple deposit) or resultant landforms characterised by a localised, very rapid type of fall in which large blocks of soil or rock literally fall over, rotating outward over a low pivot point. Portions of the original material may remain intact, although reoriented, within the resulting debris pile. Types of topples can be specified based on the dominant particle size of sediments (i.e. debris topple, soil topple, rock topple).

---

12   Slow, viscous downslope flow of water-saturated regolith

Table 30 Mass movement types

Movement types		LANDSLIDE					
	FALL Free fall, bouncing or rolling	**TOPPLE** Forward rotation over a point	**SLIDE*** Net lateral displacement along a slip face — **Rotational slide**: Lateral displacement along a concave slip face; with backward rotation. **Compound slide**: Intermediate between rotational and translational (e.g. compound rock slide)	**Translational slide**: Lateral displacement along a planar slip face; no rotation	**SPREAD** A wet layer becomes 'plastic', squeezes up and out and drags along intact blocks or beds (e.g. extrusion, liquefaction (= lateral spread))	**FLOW** The entire mass, wet or dry, moves as a viscous liquid	**COMPLEX LANDSLIDE** Combination of multiple (two or more) types of movement. No unique subtypes are recognised here; many possible. Option: name the main movement types (e.g. a complex rock spread-debris flow landslide)
Consolidated: (bedrock) Bedrock masses dominant	rockfall **(RF)**	rock topple **(RT)**	rotational rock slide **(RS)**	translational rock slide (= planar slide) (e.g. block glide) **(TS)**	rock spread **(RP)** block spread **(BP)**	rock fragment flow (e.g. rockfall avalanche) **(RL)**	
Unconsolidated coarser: Coarse fragments dominant	debris fall **(DF)**	debris topple **(DT)**	*debris slide* — rotational debris slide **(DS)**	translational debris slide **(BS)**	debris spread **(DP)**	debris avalanche (drier, steep slope) **(DA)** debris flow (wetter) **(DL)**	
Finer: Fine earth particles dominant	earth fall (= soil fall) **(EF)**	earth topple (= soil topple) **(ET)**	rotational earth slide **(ES)**	translational earth slide **(HS)**	earth spread **(EP)**	earth flow (e.g. creep, mudflow, solifluction) **(EL)**	

Dominant material

* Slides, especially rotational slides, are commonly and imprecisely called 'slumps'.
Source: Schoeneberger and Wysocki (2017), developed from Cruden and Varnes (1996).

Complex	Associated sediments (complex landslide deposit) or resultant landforms
landslide	characterised by a composite of several mass movement processes, none of
	which dominates or leaves a prevailing landform. Numerous types of complex
	landslides can be specified by naming the constituent processes evident (e.g. a
	complex earth spread – earthflow landslide).

AGGRADATION

This refers to the presence of material deposited on a pre-existing (soil) surface. Modern surface deposits can broadly be split into three categories: those that are directly anthropogenic, indirectly anthropogenic, or those that are natural. At times, it may be difficult to attribute a surface deposition as indirectly anthropogenic or natural. See also the M horizon in the Soil profile chapter.

X   *Not apparent*

0   *No aggradation*

1   *Present*

Where the origin of the aggraded material can be confidently identified, the mechanism of deposition should also be recorded in the following manner:

A   *Aeolian (wind-blown material)*

E   *Erosion (material deposited by sediment transport/runoff sourced from nearby (upslope) erosion)*

F   *Fluvial deposition (material sourced from riverine flooding)*

G   *Gravity or mass movement (material from landslides, slumping)*

P   *Pyrogenic (fire ash)*

S   *Spoil (material deposited by mechanical means, anthropogenic processes)*

INUNDATION

Inundation includes flooding from over-bank flow, inundation from local runon, overland flow and coastal events such as tidal surges and tsunamis.

Although the importance of this information is considerable, in most instances it cannot be assessed at each site. Some evidence may be available from past events – for example, accumulation of debris in trees or on fences. Otherwise information is usually obtained from local enquiry.

Frequency

Give long-term average of inundation.

0   *No inundation*

1   *Less than one occurrence per 100 years*

2   *One occurrence in between 50 and 100 years*

3   *One occurrence in between 10 and 50 years*

4   *One occurrence in between 1 and 10 years*

5   *More than one occurrence per year*

Among alluvial plains, flood plains typically fall in categories 3 and 4.

Duration (annual)
Give likely duration of an inundation event.

1   *Less than 1 day*

2   *Between 1 and 20 days*

3   *Between 20 and 120 days*

4   *More than 120 days*

Depth (annual)
Give likely maximum depth of water in an inundation event.

1   *<50 mm*

2   *50–100 mm*

3   *100–300 mm*

4   *300–1000 mm*

5   *>1000 mm*

COARSE FRAGMENTS
Coarse fragments are particles coarser than 2 mm. They include unattached rock fragments and other fragments such as charcoal and shells. Coarse fragments are distinguished from segregations of pedogenic origin in that they are not, or not considered to be, of pedogenic origin. However, coarse fragments may be altered by pedogenic processes associated with weathering.

Both coarse fragments and segregations of pedogenic origin can occur on the surface and can have a similar functional effect on land use. The description of pedogenic segregations on the soil surface should use the terms described on page 166.

Abundance of coarse fragments
The percentage is estimated by eye using the charts in Figure 18 for comparison.

0	*No coarse fragments*	0
1	*Very slightly or very few*	<2%
2	*Slightly or few*	2–10%
3	*Common*	10–20%
4	*Moderately or many*	20–50%
5	*Very or abundant*	50–90%
6	*Extremely or very abundant*	>90%

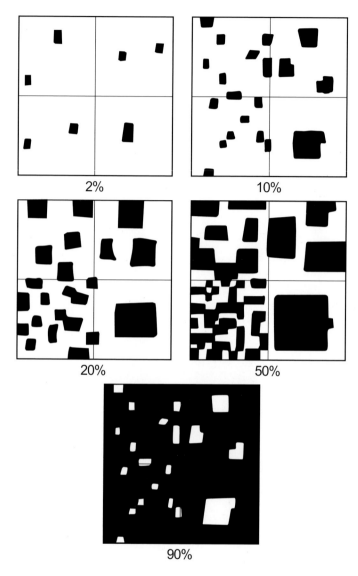

Figure 18 Chart for estimating abundance of mottles (page 137), coarse fragments (page 160) and segregations of pedogenic origin (page 166). Each quarter of any one square has the same area of black.

Size of coarse fragments

The scale adopted employs class boundaries at $2 \times 10^{n/2}$ mm, where n is an integer. It is compatible with both the International Scheme referred to in the field texture section (page 138) and the grain size criteria for substrate materials (page 178).

The terms used to describe size apply to fragments *of any shape*. The *average maximum dimension* of fragments is used to determine the class interval. Figure 19 can be used as a field guide for determining the size of particles.

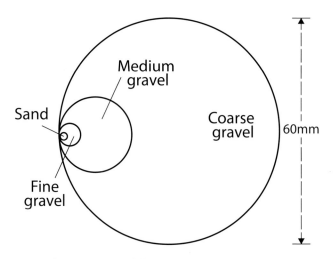

Figure 19   Relative size of sand and gravel (from FAO 2006).

1	*Fine gravelly[13] or small pebbles*	2–6 mm
2	*Medium gravelly or medium pebbles*	6–20 mm
3	*Coarse gravelly or large pebbles*	20–60 mm
4	*Cobbly or cobbles*	60–200 mm
5	*Stony or stones*	200–600 mm
6	*Bouldery or boulders*	600–2000 mm
7	*Large boulders*	>2000 mm

Shape of coarse fragments

Give shape using Figure 20 as a visual guide and using the terms below.

A	*Angular*
S	*Subangular*
U	*Subrounded*
R	*Rounded*
AT	*Angular tabular*
ST	*Subangular tabular*
UT	*Subrounded tabular*
RT	*Rounded tabular*
AP	*Angular platy*
SP	*Subangular platy*
UP	*Subrounded platy*
RP	*Rounded platy*

---

13   Note that in preparing soils for laboratory analysis, the greater than 2 mm size fraction is commonly recorded as 'gravel'.

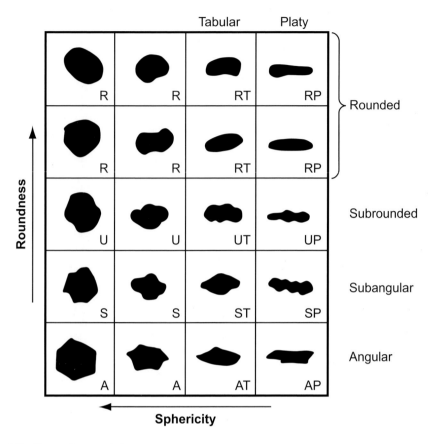

Figure 20   Coarse fragment shapes.

Lithology of coarse fragments

M   *Same as substrate material (page 126)*

R   *Same as rock outcrop*

Where the lithology of coarse fragments is different from that of the substrate material and/or rock outcrop, describe it as for lithology of substrate material (see Table 35, page 182). Some coarse fragments are commonly encountered that are not listed in Table 35. These include:

CC   *Charcoal*

IS   *Ironstone (where not considered of pedogenic origin)*

OW   *Opalised wood*

PU   *Pumice*

SS   *Shells*

OT   *Other*

Alteration of coarse fragments

Alteration of a lithology can occur through a variety of processes, often leading to enrichment with a particular element, such as silica or iron. The alteration of coarse fragments should be described as per page 189.

Strength of coarse fragments

Same as for 'Strength of substrate material' (see page 180).

Magnetic attributes of coarse fragments

Same as for 'Magnetic attributes of segregations' (see page 168).

ROCK OUTCROP

This refers to any exposed area of rock that is inferred to be continuous with underlying bedrock.

Abundance of rock outcrop

0	*No rock outcrop*	No bedrock exposed.
1	*Very slightly rocky*	<2% bedrock exposed.
2	*Slightly rocky*	2–10% bedrock exposed.
3	*Rocky*	10–20% bedrock exposed.
4	*Very rocky*	20–50% bedrock exposed.
5	*Rockland*	>50% bedrock exposed.

Distance between rock outcrop

1	*>50 m*
2	*20–50 m*
3	*5–20 m*
4	*2–5 m*
5	*<2 m*

Lithology of rock outcrop

Record lithology of rock outcrop as for lithology of substrate material (see page 182, Table 35).

DEPTH OF/TO FREE WATER

Give depth (in metres) of/to free water at the site of soil observation. Free water is recorded as a positive value (height above the soil surface) or a negative value (depth below the soil surface), excluding litter and living vegetation. Measurements should be cognisant of the precision that is realistically achievable. Prefix the depth above or below the soil surface as follows:

+	*Above soil surface*
–	*Below soil surface*

If there is no free water, record:

Z *No free water*

CONDITION OF SURFACE SOIL WHEN DRY

The soil surface has a characteristic appearance when dry. Because surface conditions are often relevant to the use of the soil and indicative of particular kinds of soil, every effort should be made to observe the surface condition in the dry state. Terms broadly concern the consistence of the soil surface (e.g. loose, soft) and the nature of the soil surface (e.g. self-mulching, surface flake). Some terms describing the nature of the soil surface are mutually exclusive (e.g. a surface cannot be snuffy and hardsetting). Other terms invariably occur together (e.g. cracking and self-mulching). Terms describing nature can be used in conjunction with consistence terms (e.g. soft, snuffy surface; firm surface crust). A more detailed description of crusting and hardsetting soils is provided in So *et al.* (1994).

L	*Loose*	Incoherent[14] mass of individual particles or aggregates. Surface easily disturbed by pressure of forefinger.
S	*Soft*	Coherent[15] mass of individual particles or aggregates. Surface easily disturbed by pressure of forefinger.
F	*Firm*	Coherent mass of individual particles or aggregates. Surface disturbed or indented by moderate pressure of forefinger.
H	*Hardsetting*	Compact, hard, apparently apedal condition forms on drying but softens on wetting. When dry, the material is hard below any surface crust or flake that may occur, and is not disturbed or indented by pressure of forefinger.
G	*Cracking*	Cracks at least 5 mm wide and extending upwards to the surface or to the base of any plough layer or thin (<0.03 m) surface horizon.
W	*Weakly self-mulching*	Weakly pedal, often partially crusty surface mulch that forms on wetting and drying. Peds commonly less than 5 mm in least dimension.
M	*Self-mulching*	Strongly pedal loose surface mulch forms on wetting and drying. Peds commonly less than 5 mm in least dimension.
X	*Surface flake*	Thin, massive surface layer, usually less than 10 mm thick, which on drying separates from, and can be readily lifted off, the soil below. It usually consists mainly of dispersed clay, and may become increasingly fragile as the soil dries.
C	*Surface crust*	Distinct surface layer, often laminated, ranging in thickness from a few millimetres to a few tens of millimetres, which is hard and brittle when dry and cannot be readily separated from, and lifted off, the underlying soil material.

---

14 Incoherent means that less than two-thirds of the soil material, whether composed of peds or not, will remain united at the given moisture state without very small force (force 1, see 'Consistence' on page 161) having been applied.

15 Coherent means that two-thirds or more of the soil material, whether composed of peds or not, will remain united at the given moisture state unless force is applied.

N	*Snuffy*	Loose, powdery or pulverescent finely structured surface with low bulk density, usually associated with iron-rich soils (Ferrosols). It is often strongly acidic and water repellent. Not to be confused with a 'dusty' soil surface associated with high silt fraction in some soils.
V	*Gravel pavement*	Thin, more or less continuous surface cover of pebbles (<60 mm in size), typically strongly altered by weathering and wind erosion (desert varnish).
Y	*Cryptogam surface*	Thin, more or less continuous crust of biologically stabilised soil material usually due to algae, liverworts and mosses.
T	*Trampled*	Soil that has been extensively trampled under dry conditions by hoofed animals.
P	*Poached*	Soil that has been extensively trampled under wet conditions by hoofed animals.
R	*Recently cultivated*	Effect of cultivation is obvious.
Z	*Saline*	Surface has visible salt, or salinity is evident from the absence or nature of the vegetation or from soil consistence. These conditions are characterised by their notable difference from adjacent non-saline areas.
O	*Other*	

RUNOFF

Runoff is the relative rate at which water runs off the soil surface. It is largely determined by slope, surface cover and soil infiltration rate.

0	*No runoff*	
1	*Very slow*	Free water on surface for long periods, or water enters soil immediately. Soils usually either level to nearly level or loose and porous.
2	*Slow*	Free water on surface for significant periods, or water enters soil relatively rapidly. Soils usually either nearly level to gently sloping or relatively porous.
3	*Moderately rapid*	Free water on surface for short periods only; moderate proportion of water enters soil. Soils usually gently sloping to moderately inclined.
4	*Rapid*	Large proportion of water runs off; small proportion enters soil. Water runs off nearly as fast as it is added. Soils usually have moderate to steep slopes and low infiltration rates.
5	*Very rapid*	Very large proportion of water runs off; very small proportion enters soil. Water runs off as fast as it is added. Soils usually have steep to very steep slopes and low infiltration rates.

SOIL PROFILE

RC McDonald, RF Isbell and AJW Biggs

There are numerous definitions of *soil* in use, but all contain some key concepts: soil is the unconsolidated to semi-consolidated mineral matter and transformed/decomposed organic matter that lies on the earth's surface; it is the result of various pedogenetic processes acting upon mineral and organic parent materials; and it supports (or can support) the growth of plants. If too strictly interpreted, some definitions could exclude materials such as bare sand dunes or lakebeds from being considered as soil. However, such materials are included in the definition of soil, despite at times an absence of vegetation and/or limited evidence of pedogenic processes – noting that limited evidence does not mean that pedogenic processes are absent, it is simply that such processes may not be overtly visible.

There is an important distinction between *soil* (the material) and a *soil profile*. A soil profile is a vertical section of soil from the surface through all its *horizons/layers* to bedrock, other consolidated *substrate* material or to a selected depth in unconsolidated material. A soil profile is both a conceptual entity and a physical entity that is described in a *soil profile description*, although the description may only portray a portion of the entity. A soil profile is part of the *regolith* (see the Substrate chapter for further detail). Soil description typically ceases once weathered rock is encountered, whereas in regolith science, weathered rock is a zone of key interest.

Traditional representations of a conceptual soil profile often include some non-soil layers, such as rock and litter, due to their relevance to pedogenesis and soil properties (a *layer* is comprised of material that is not soil, cf. a *horizon*, which is comprised of soil). Thus, for the purposes of soil profile description, it is common practice to include all soil horizons and those immediately adjacent layers of non-soil material (where they exist and can be described). Consequently, a soil profile description may extend beyond the bounds of the soil profile *per se*.

The soil profile should not be confused with the *solum*, in which the surface and subsoil layers have undergone the same soil forming conditions (Isbell and NCST 2021). The solum consists of A and B horizons, their transitional horizons, and P and O horizons – that is, it is a portion of the soil profile. A solum does not have a maximum or a minimum thickness.

A soil profile can be seen as an individual (Macvicar 1969, page 143; Northcote 1979, page 22) that is described by giving a single value to each property. This is distinct from the pedon (Soil Survey Staff 1975), soil series or soil profile class (Powell 2008), in which the soil body is described by a range of values for each property. Considering the variability inherent in soils, ideally a soil description would give a range of values for each property recorded in each of the three dimensions in each horizon. In practice this is not possible, as the pedologist can describe factually only the very small parts of the soil body actually seen. Most soil descriptions are necessarily given with a single value for each property described.

TYPE OF SOIL OBSERVATION

The soil profile may be described/sampled via the following methods of exposure (listed in order of preference):

P	*Soil pit*
E	*Existing (near) vertical exposure*
C	*Relatively undisturbed soil core*
S	*Shallow soil pit (<0.3 m)*
A	*Auger boring*
U	*Vacuum soil sampling*
Y	*Surface observation*

To characterise a soil profile fully, it should be examined to the depth of the parent material or other consolidated material. However, because soil depth varies between very wide limits and because soils are examined for a wide variety of purposes, the depth of examination in practice frequently may not exceed 1.5–2 m. Aspects of soil profile description, such as horizon nomenclature and consequent classification, can be significantly influenced by the depth of exposure and description – this is a largely unavoidable problem that should be borne in mind when exposing, describing and classifying a soil profile.

Use of augers and vacuum samplers can create significant disruption to the soil and the ability to accurately describe many attributes is significantly reduced. Vacuum samplers are efficient for sampling loose sandy soils but not recommended for soil profile description.

If an R layer or pan is not encountered, it is beneficial to record contextual information regarding the lowest depth of exposure of the soil profile that has been achieved. The depth achieved could be due to an inability to extract soil (mechanical resistance, too loose, too wet), or equipment limitations. The reason provided is typically related to the method of exposure/extraction of the soil, thus it is essential that the type of soil observation is also recorded.

E	*Extent of equipment*	Depth of exposure dictated by length of equipment used (e.g. 1.5 m auger).
H	*Mechanical resistance (hard soil)*	Resistance from high bulk density or high soil strength prevents further penetration (in the context of the equipment being used).

L	*Too loose*	Lack of soil cohesion limits retrieval of materials (and the soil is dry).
M	*Mechanical resistance (hard objects)*	Segregations, coarse fragments or unknown hard object prevent further penetration.
N	*No reason*	Neither equipment nor soil factors have limited the depth of exposure.
W	*Too wet*	Lack of soil cohesion limits retrieval of materials (and the soil is saturated).

HORIZONS AND ASSOCIATED LAYERS

A soil horizon is a layer of soil material, approximately parallel to the land surface, with morphological properties different from horizons/layers below and/or above it. Tongues of material from one horizon may penetrate into adjacent horizons.

Horizon notation in Australia differs in part from that often adopted overseas, noting that there is considerable variation in nomenclature in the international arena. Emphasis is on factual objective notation rather than assumed genesis, as genetic implications are often uncertain and difficult to establish. Thus the notation E, indicating eluvial horizon (International Society of Soil Science 1967), has not been adopted, even though this has been implemented by a number of overseas organisations (e.g. Hodgson 1974; FAO 2006; Soil Science Division Staff 2017; IUSS Working Group 2022). Harms (2023) provides a discussion of horizon nomenclature in Australia and internationally.

As horizon notation is deduced from the profile description data (Northcote 1979), it is recorded *after* the profile is described – it is a form of classification of profile attributes. However, for the purposes of explanation within this chapter, the description of horizons is provided prior to the description of attributes. The importance of accurate description prior to classification (either of horizon nomenclature or the soil profile) cannot be over-stated. Confirmation of nomenclature may be assisted by laboratory data, but horizons should be named in the field as completely as possible.

The layers and horizons that are typically encountered in soil profiles formed on weathering rock are depicted in Figure 21. While they are shown in the normal order encountered, *there is no requirement for all horizons to exist in a given soil profile*. Unless an A1 horizon has been removed by erosion or other means, it is typically the uppermost mineral horizon of a soil profile.

Organic layers and horizons

The three organic layers/horizons (L, O and P) have similar characteristics, in that they are derived from biomass production on or in the soil, and the rate of biomass production exceeds the rate of decomposition (hence the accumulation of organic material[16]) in either the short or long term, in either past or present conditions. Litter (the L layer) is the most

---

16   Note that the term *organic material* as used in this text is generic and is not the definition used within the Australian Soil Classification

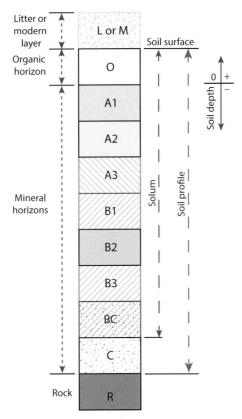

Figure 21 Horizons and layers that can exist in, and adjacent to, a soil profile formed on weathering rock. P and D horizons are not shown.

easily defined. In the case of O and P horizons, however, both are comprised of partially or completely decomposed organic matter and have many other similar properties, such as very low bulk density. The key differentiation relates to their conditions of formation, in particular the frequency of saturation – O horizons forming in what are essentially 'dry' conditions and P horizons forming in 'wet' conditions. There are other, more complex differences related to geochemistry and species of vegetation present during formation.

L *layer*[17]

This is an incoherent (loose), organic (non-mineral) layer accumulated in predominantly unsaturated conditions on the surface of a mineral or organic soil. It is derived via fall of material from standing vegetation and is dominated by undecomposed organic materials (litter). Organic materials are typically produced *in situ*, but may also be transported in. It is easily distinguished from underlying O horizons (where they exist). While an L layer is not soil material (and therefore not a horizon), it is often described with the soil profile because of its important contribution to soil processes. The L layer is commonly

---

17   This was denoted as the O1 horizon in previous editions of the Field Handbook

lost in Australia through grazing, fire and erosion and may be seasonal in its formation/presence.

| L | *Layer* | Consists of undecomposed organic debris, usually dominated by leaves, bark and twigs, accumulated on the soil surface in predominantly unsaturated, aerobic conditions via debris fall. The original form of the debris can be recognised with the naked eye. |

O *horizon*[18]

This is an incoherent to somewhat coherent surface horizon dominated by organic (non-mineral) materials in varying stages of decomposition that have accumulated in predominantly unsaturated (dry) conditions on the surface of a mineral or organic soil. Organic materials may be produced *in situ* or transported in by natural or anthropogenic processes. Typically an O horizon is derived from the decomposition of an L layer, but may also be derived from decomposition of *in situ* biomass (e.g. fibrous roots). There is often (but not always) a visible discontinuity between an O horizon and underlying mineral horizons – that is, the organic material is not fully integrated with the underlying horizon and is easily separated from the surface of the mineral soil.

An O is regarded as soil material (and therefore a horizon), because the organic matter has become part of the pedogenic domain – that is, it is partly decomposed and may be mixed with mineral materials (clay, silt, sand, segregations, coarse fragments). It is possible to have an L or O individually or in combination. An O horizon is most commonly removed via decomposition, erosion or fire and may be seasonal in its formation/presence.

| O | *Horizon* | Visually distinct horizon consisting of organic debris in various stages of decomposition. The original form of most of the debris cannot be recognised with the naked eye. The horizon may also contain some mineral material. |

P *horizons*

These are typically coherent, predominantly organic (non-mineral) horizons that have developed in wet conditions where the rate of accumulation of organic material exceeds the rate of decomposition. Those conditions may have been historical and not currently evident. Influencing environmental factors leading to their formation include, but are not limited to, prolonged waterlogging, temperature, redox status, nutrient status, species of plants and microbial activity. In general, one or more factors result in very low rates of decomposition/loss of organic materials. The source of organic materials is generally *in situ* biomass (plant material) produced above and/or below-ground. P horizons may be found uppermost or within the profile (e.g. as a buried horizon). When such horizons are buried, they may be designated in a manner similar to designations for buried mineral soils (see page 133) – for example, 2Pαb, 3Pθb, etc.

P horizons have long been known colloquially as *peat*. Recent formalisation of the definition of peat in Australia (see Appendix 2) means that a P horizon may not always qualify as peat – it

---

18   This was denoted as the O2 horizon in previous editions of the Field Handbook

may be comprised of peat or *peat-like* materials. The determination of a horizon as a P relies solely upon field indicators, whereas the determination of materials as peat relies in part upon laboratory analyses.

Three subdivisions are recognised,[19] depending on stages of decomposition.

Pα   *Horizon*   Consists primarily of undecomposed or weakly decomposed organic material (fibric organic material). Plant remains are distinct and readily identifiable. If wet, yields clear to weakly turbid water when squeezed; no material escapes between fingers.

Pγ   *Horizon*   Consists primarily of moderately to well decomposed organic material (hemic organic material). Plant remains vary from being recognisable to unrecognisable. Yields strongly turbid to muddy water when squeezed; amount of material escaping between fingers ranges from none up to one-third; residue is pasty.

Pθ   *Horizon*   Consists primarily of strongly to completely decomposed organic material (sapric organic material). Plant remains vary from being indistinct to unrecognisable.

P horizons often occur in sequences reflecting varying deposition/formation conditions. Nomenclature is applied in a similar manner to mineral horizons (e.g. Pα1, Pα2, Pγ1, Pγ2). In FAO (2006), peat horizons are referred to as H horizons.

Peat horizons, while coherent, are subject to modification via changes to their hydrological environment. Examples include, but are not limited to, land drainage, droughts, wildfires and climate change. P horizons, due to their organic characteristics, are extremely susceptible to wildfire when dry, resulting in partial, or in some situations, complete loss and oxidation.

Mineral layers and horizons

The following mineral layers and horizons may be described in/on/under a soil profile. Subdivision (e.g. A1, B21) is not necessary if only one instance of the horizon exists or there is considerable doubt as to the appropriate nomenclature. However, it is recommended that all efforts are made to ensure that horizons are designated in as detailed a manner as possible in the field, as this can have significant impacts on subsequent use of data – for example, simply designating a horizon as A may create uncertainty as to whether it is an A1 or A2 horizon.

M *layers*

These layers are the result of recent deposition of material over an existing soil profile. They are a characteristic feature of the site, are the result of natural or anthropogenic processes, and are unrelated to the underlying material (constituting a lithologic discontinuity – see page 133). An M layer has no evidence of post-deposition pedological development (no organic matter accumulation, structural development etc.) and hence does not meet

---

19   In previous editions, only two subdivisions were recognised (P1 and P2). P1 corresponds to Pα. P2 is now subdivided into Pγ and Pθ.

the definition of a horizon. It can have a similar nature to a *C layer* (lack of pedogenic development), but can confidently be determined to be the result of recent, surficial depositional processes (aggradation). An M layer should be recognisable and contiguous, with a minimum thickness of 0.01 m. Its thickness is measured as a positive value **above** the soil profile (as for L and O).

An M layer also encompasses ash materials from fire, as this is modern depositional material. The composition of ash is influenced by the temperature of the fire and fuel source, among other things (Bodí *et al.* 2014). The *ash material* and *carbic* concepts on the soil surface, as described in *The Australian Soil Classification* (Isbell and NCST 2021), are encompassed within the definition of an M layer. Ash material found within a soil profile would be regarded as a C layer.

The nature and definition of an M layer is such that its presence does not invoke the use of a prefix (2) or suffix (b for buried) for the immediately underlying soil profile (as for L and O[20]).

M	*Layer*	Visually distinct layer lacking pedological development, deposited on the soil surface. It results from modern deposition through natural or anthropogenic processes.

A *horizons*

These are horizons either consisting of:

- one or more surface mineral horizons with some organic accumulation, and usually darker in colour than the underlying horizons; or
- surface and subsurface horizons that are lighter in colour but have a lower content of silicate clay and/or sesquioxides than the underlying horizons.

A1	*Horizon*	Mineral horizon at or near the soil surface with some accumulation of humified organic matter; **usually** darker in colour than underlying horizons. Zone of maximum biological activity (although this may be episodic). It may be divided into sub-horizons and of these, the A11 horizon is usually the more organic, or darker coloured uppermost portion. The A12 differs in colour, structure or texture from the A11, usually being lighter in colour. It is not pale enough, however, to qualify as an A2 horizon. The A1 may be further divided into sub-horizons if necessary – for example, A13, A14 – although this is rarely required.
A2	*Horizon*	Mineral horizon having, either alone or in combination, *less* organic matter, sesquioxides or silicate clay than immediately adjacent horizons. It is usually differentiated from the A1 horizon by its paler colour, that is, by having a colour **value at least one unit higher** and less organic matter. It is usually differentiated from the B horizon by having colour **value at least one unit higher** and **chroma at least two units lower**, by coarser texture or by a combination of these attributes.
A3	*Horizon*	*Transitional horizon* between A and B, which, while containing traits of both, has properties more characteristic of the overlying A horizon.

---

20   See also the section on recording soil depth

Refer also to the section on determining A and B horizons in cracking clays (page 131).

B *horizons*

These are horizons consisting of one or more mineral soil horizons characterised by one or more of the following:

- a concentration of silicate clay, iron, aluminium, manganese, carbonate, salt(s), organic material or several of these
- structure and/or consistence unlike that of the A horizons above, or of any horizons immediately below
- stronger colours, usually expressed as higher chroma and/or redder hue than those of the A horizons above or those of the horizons below.

B1	*Horizon*	*Transitional horizon* between A and B, which, while containing traits of both, has properties more characteristic of the underlying B2 horizon.
B2	*Horizon*	Horizon in which the dominant feature is one or more of the following:

> - an illuvial, residual or other concentration of silicate clay, iron, aluminium, manganese, carbonate, gypsum or other salts, or humus, either alone or in combination
> - maximum development of pedologic organisation[21] as evidenced by a different structure and/or consistence, and/or stronger colours than the A horizons above or any horizon immediately below.

> It may be divided into sub-horizons – for example, B21, B22, B23.

B3	*Horizon*	Transitional horizon between B and C, or other sub-solum material, in which properties characteristic of an overlying B2 dominate, but intergrade to those of the underlying material.

Refer also to the section on determining A and B horizons in cracking clays (page 131) and the section concerning lithologic discontinuities (page 133).

C *layer*

These are layers (typically below the solum) of consolidated or unconsolidated material only weakly affected by pedogenic processes such as weathering, and are either like or unlike the material from which the solum presumably formed. The C layer lacks properties characteristic of M, L, O, P, A, B or D layers/horizons. It is recognised by its lack of pedological organisation and/or the presence of geologic organisation frequently expressed as sedimentary laminae or as ghost rock structure as found in saprolite. C layers include consolidated rock and sediments that, when moist, can be dug with hand tools. Rock strength is generally weak or weaker. Because of their nature, C layers may be described as detailed in this chapter or as substrate (see page 177). C layers also include unconsolidated sediments lacking in pedological development

---

21 Pedologic organisation is a broad term used to include all changes in soil material resulting from the effect of physical, chemical and biologic processes i.e. pedogenesis. Results of these processes include horizonation, colour differences, presence of pedality, pedogenic segregations and texture and/or consistence changes.

(e.g. sand or marine mud). These are distinguished from 'weathered rock' C layers by use of the suffix r for the latter (see page 132).

C   *Layer*   Consolidated or unconsolidated layer lacking pedological organisation or properties characteristic of other layers/horizons (M, L, O, P, A, B, D). Includes consolidated rock and sediments (saprolite) that can be dug with hand tools.

D *horizons*

These are any **soil material** below the solum that: a) possesses pedological organisation, b) is unlike the solum in its general character, c) is not a C layer, and d) cannot otherwise be given reliable horizon designation as described in the sections on Lithologic discontinuities or Buried soils (see pages 133 and 208).

D   *Horizon*   Horizon displaying pedological organisation that is unlike the overlying solum, is not a C layer and cannot be reliably designated as another horizon.

R *layer*

This layer consists of continuous masses (not boulders) of moderately strong to very strong rock (excluding pans, page 163) such as bedrock. It is not soil material and is described using terminology in the Substrate chapter. The designation of an R layer is for convenience only, as it is strictly not part of the soil profile.

R   *Layer*   Continuous masses of moderately strong to very strong rock (excluding pans). Not comprised of soil material.

Transitional horizons

Three kinds of transitional horizons are distinguished:

- transitional horizons between A and B or B and C horizons/layers, in which the properties of one horizon are dominant (these have been described above in their relevant sections as A3, B1 and B3 horizons)
- transitional (combination) horizons that have subordinate properties of both horizons but are not dominated by properties characteristic of either horizon, and it is not possible to designate as A3, B1 or B3. For example, AB, AC, BC horizons.
- transitional (combination) horizons in which distinct parts of two different horizons are clearly visible, indicated by capital letters separated by a virgule (/) as in A/B, B/A, A/C, B/C, C/B. Most of the individual parts of at least one of the components are surrounded by the other. The first symbol is that of the horizon that makes up the greater volume.

The difference between a B3 and a BC horizon is typically the most difficult, as there is often a continuum present (Figure 22). In a B3, soil material is dominant and some characteristics of the underlying horizon (often a Cr) are recognisable, but subordinate (e.g. presence of many weathered substrate rock fragments). In a BC horizon, the proportion of soil material is lesser and it is difficult to distinguish dominant characteristics. As a guide,

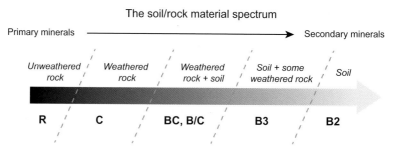

Figure 22 The conceptual material spectrum for a soil forming on rock.

it is typically easy to obtain a field texture from a B3 horizon, whereas it is more difficult to do so in a BC horizon.

BC   Horizon in which properties of both B and C horizons/layers are found, in a manner that the individual components of the two horizons are not easily separated.

B/C   Horizon in which properties of both B and C horizons/layers are found as discretely recognisable and separate components (if the properties of the C dominate it is designated as a C/B horizon).

Bleached horizons

Some horizons are white, near white or much paler than adjacent horizons. Bleached horizons most commonly occur as A2 horizons but *are not restricted to them*. Bleached horizons are defined in terms of the Munsell system of colour notation for *dry* soil:

* for all hues, value 7 or greater with chroma 4 or less
* where adjacent horizons have hues 5YR or redder, value 6 or greater with chroma 4 or less.

Two kinds of bleached horizons are recognised:

Conspicuously bleached: 80% or more of the horizon is bleached (suffix e)
Sporadically bleached: bleach occurs (suffix j) as:
* irregularly through the horizon, or
* blotches, often less than 6 mm thick, at the interface of horizons, most commonly A and B horizons, or
* nests of bleached grains of soil material at the interface of horizons, most commonly A and B horizons, when no other evidence of a bleached horizon may occur.

There are a number of specific instances of bleached horizons that can be difficult to deal with. One is where the bleach occurs at the interface between the A and B horizons in texture contrast soils, particularly those with large columnar structure. In such soils, the bleach may extend some distance down the cracks between columns and be either continuous or discontinuous (sporadic). In this case, the bleach is not part of the B horizon, but its presence may be indicated via description of the nature of the horizon boundary as tongued. In some texture contrast soils, a bleach may extend into the upper 5–10 mm of the B horizon and/or occur as blotches (as indicated above). Care is required to discriminate if this is the case (likely to be a very thin B1e), or the bleach is only present as an A horizon (A2e). Sporadic bleaches are often very thin (<0.01 m). In such circumstances, convention is to describe the horizon as 0.01 m thick.

In arid and semi-arid areas, where A1 horizon development is minimal, bleached horizons can occur at the soil surface, either as a result of loss of the A1 horizon and exposure of a bleached A2, or because the soil is inherently pale/bleached and has no A1 horizon evident by colour – for example, a bleached windblown sand or some very silty soils.

Horizons in cracking clays

The A and B horizons in cracking clays are defined and recognised on the basis of structure rather than colour (McDonald 1977).

The A horizon in cracking clays may be structured or massive. The structural A horizon is the granular, subangular blocky, angular blocky or polyhedral surface horizon(s) where ped faces are not accommodated and have irregular coarse voids between them. This is exemplified in soils with a self-mulching surface. The A horizon structure is unstable, in the sense that relatively rapid wetting and drying continually create new peds and voids.

The B horizon is the coarse prismatic and/or angular blocky and/or lenticular horizon, where ped faces are all accommodated and usually have only narrow planar voids between them. The B horizon structure is comparatively stable because of relatively slow wetting and drying. The B horizon boundary will usually occur within 200 mm of the surface, but the structural change must be confirmed using a spade or exposure to examine the soil. The boundary may often be gradual or diffuse, but can be clear or abrupt.

Subdivision of horizons

All horizon subdivisions are numbered consecutively from the top of each horizon downward, as in A11, A12, A13, A21, A22, A23, B21, B22, B23, B31, B32, B33, C1, C2, C3, D1, D2, D3 etc.

When required, a numeric suffix always precedes the alphabetical suffix *except* with the alphabetical suffix 'p' where the number always follows the letter (i.e. Ap1, Ap2). As many suffixes are used as necessary and are typically written in alphabetical order (e.g. B2hs), with the exception of the b suffix, which is applied last.

The above horizon nomenclature will cover most soils, but there will be instances where there are buried soils (pedologic discontinuities, page 133) and also where a profile has formed in what are obviously different parent materials (lithologic discontinuities, page 133).

Horizon suffixes

a   Used for horizons with accumulation of jarosite.

b[22]   Used for soil horizons that display features developed prior to burial. The suffix is written last (e.g. 2B2b).

---

22   This suffix should only be used according to the definition of buried soils on page 208.

c[23] Used for horizons/layers with accumulation of concretions or nodules of iron and/ or aluminium and/or manganese (as in B2c, B21c).

cc Used for horizons with >50% concretions or nodules of iron and/or aluminium and/ or manganese by volume (as in B2cc, B21cc).

d Used for densipans – very fine, sandy, earthy pan (see page 165).

e Used for conspicuously bleached horizons (e.g. A2e).

f Used when faunal accumulation, such as worm casts, dominates certain A1 horizons (e.g. A1f in some soils under rainforest).

g Indicates strong gleying, as in B2g. Gleying is indicative of permanent or periodic intense reduction due to wetness; it is characterised by greyish, bluish or greenish colours, generally of low chroma. Mottling may be prominent; mottles may have reddish hues and higher chromas if oxidising conditions occur periodically. Roots may have rusty or yellowish outlines; hence horizons such as A1g can occur.

h Used where horizons contain accumulation of amorphous, organic matter– aluminium complexes in which iron contents are very low. The dominantly organic matter–aluminium complexes occur as discrete pellets between clean sand grains or completely fill the voids; occasionally they may coat sand grains. Such horizons may be soft or cemented and form the characteristic B horizon of poorly drained soils known as podzols or Podosols (see also ortstein pans, page 165).

i Used for horizons with vesicular pores, most commonly A horizons.

j Used for sporadically bleached horizons (e.g. A2j).

k[23] Used for horizons with accumulation of carbonates, commonly calcium carbonate (as in B2k, B21k).

kk Used for horizons with >50% secondary carbonates by volume (as in B2kk, B21kk).

m Used for horizons with strong cementation or induration. It is confined to irreversibly cemented horizons that are essentially continuous (more than 90%), although they may be fractured (see page 163).

n Used for horizons that have changed properties (primarily changed to an orange/red colour) as a result of heat (fire).

p Used for horizons where ploughing, tillage practices or other disturbance by humans has occurred (e.g. deep ripping). This suffix is typically used only with A (as in Ap). Where the plough layer clearly includes what was once B horizon and it is no longer possible to infer with any reliability what the texture and depth of the A horizon was, the plough layer is designated Ap. An Ap horizon may be subdivided into sub-horizons – for example, Ap1, Ap2. *Note*: An Ap2 horizon is not the same as an A2 horizon but a subdivision equivalent to A12.

q Used for horizons with accumulation of secondary silica. If silica cementation is continuous or nearly continuous, 'qm' is used.

r Used for horizons with layers of weathered rock (including saprolite) that, although consolidated, can be dug with hand tools.

---

23 Note that the useage of these suffixes has changed from previous editions of the Field Handbook.

s   Used for horizons with an accumulation of sesquioxide–organic matter complexes in which iron is dominant relative to aluminium. These complexes coat sand grains, occur as discrete pellets, or, with moderate amounts of iron, may fill voids forming cemented patches. The content of organic matter is variable and its distribution is often irregular. The suffix 's' is often used in combination with 'h' (as in Bhs) where both organic and iron components are significant; Bs or Bhs horizons may be soft or hard and form the characteristic B horizon of free-draining Podosols.

t[23]   Used for horizons with accumulation of silicate clay (from German *ton*, clay) relative to the preceding horizon. Different mechanisms (such as illuviation, formation *in situ*) may be responsible for the clay accumulation, but these may be difficult to confirm. The suffix is applied to the lower of two adjacent horizons when there is an increase of one or more field texture classes between those two horizons.

u   Used for horizons showing evidence of strong reduction.

v   Used for horizons with vertic properties (slickensides, lenticular structure), most commonly B horizons.

w   Used where development of colour and/or structure, or both, in the B horizon is observed, with little or no accumulation of sesquioxide–organic matter complexes.

x   Used for fragipans or earthy pans. Horizon(s) with high bulk density relative to the horizon above, seemingly cemented when dry, but showing a moderate to weak cementation when moist (see page 164).

y[23]   Used for horizons with <50% gypsum (calcium sulphate) segregations by volume (as in B2y, B21y).

yy   Used for horizons with >50% gypsum segregations by volume (as in B2yy, B21yy).

z   Used for horizons with evident accumulation of salts more soluble than calcium sulphate and calcium carbonate.

?   Used where doubt is associated with the nomenclature of the horizon, with the query following the horizon notation (e.g. 'D?').

PEDOLOGIC AND LITHOLOGIC DISCONTINUITIES

A pedologic discontinuity is a vertical change between two sequential soil profiles, or parts thereof (i.e. there is more than one soil profile/material evident in the section observed). It may or may not be associated with a *lithologic discontinuity,* in which there is an obvious contrast (vertical change) in the lithology of the materials – either between horizons or between the soil profile and the underlying lithology. Pedologic discontinuities also include *buried soil* profiles – these are typically associated with a lithologic discontinuity. However, not all pedologic discontinuities are associated with a lithologic discontinuity and not all lithologic discontinuities lead to a pedologic discontinuity. A pedologic discontinuity can also occur as a *polygenetic profile*, in which more than one soil profile has formed *in situ*, either at the same time or as a sequence over time (see Appendix 3).

Below the point of pedologic or lithologic discontinuity, subsequent horizons/layers are given a numeric prefix, numbered downward in sequential order. The same numeric prefix is retained for each horizon if they are deemed to be associated, provided there is no subsequent pedologic or lithologic discontinuity. Alluvial soil profiles involving a fining-up sequence are not regarded as a lithologic discontinuity, as they are by definition a continual sequence of deposits of increasing clay content. Examples of discontinuities, fining-up sequences and their associated nomenclature are provided in Appendix 3.

DEPTH OF HORIZONS/LAYERS

The *upper and lower depths*, in metres, of each horizon or layer (P, A, B, C, D, R) are measured from the soil surface, excluding L and M layers and the O horizon, as these are often transient. L, M and O layer/horizon depths are measured above the mineral soil surface – for example, L 0.12–0.10 m, O 0.10–0.0 m (Figure 21).

The allocation of horizon/layer depths is often the first attribute recorded for a profile, but initial depths should always be confirmed during the description of attributes – for example, pH may change down a profile without any obvious morphological change. A scaled profile diagram may be useful where horizons are irregular.

Eroded soils

Description of eroded soils is not ideal practice in the context of representing undisturbed landscapes. It is, however, unavoidable in some landscapes, and a deliberate activity in certain types of work. Consequently, a consistent approach is required. The profile description should start at 0, as this depth is defined as the soil surface, irrespective of what mineral horizon is uppermost in the profile. A note should be made of the depth of material lost, if this can be confidently determined.

Allocation of horizons in eroded soils requires some consideration and understanding of the profile as it is likely to have existed prior to truncation. This is preferably achieved by exposure of an un-eroded profile of the same soil within close proximity, but at times it will require expert judgement.

If the observer is confident, then the horizons may be described as they functionally existed in the profile prior to truncation. This is particularly relevant to profiles in which the A1 horizon has been lost and the uppermost horizon is an A2 – the definition of an A2 normally requires consideration of the horizon above and below the horizon in question. Similarly, the observer may be confident that the uppermost horizon is a B2. If there is a lack of confidence for any reason, a conservative approach would be to find an alternate profile or refrain from allocating horizon nomenclature.

DEPTH TO R LAYER OR STRONGLY CEMENTED PAN

One of the most important features of a soil is its depth or thickness, but it is frequently difficult to determine the lower boundary of a soil profile. For many purposes, depth of soil is considered to be synonymous with the rooting depth of plants; however, as plant species vary widely in their rooting depth, it is not always a suitable criterion. Thickness of solum (P + O + A + B horizons) is a measure that is useful in many soils, although it may be difficult in some soils to distinguish B from C or D horizons. In soils underlain by unconsolidated mineral materials, the unconsolidated materials may be included in soil depth calculations.

The purpose of recording this depth is to include in the soil description those materials that may be relatively easily moved by earth-moving equipment or may be relatively easily penetrated by roots. Strong or very strong rock or pan that may require ripping or blasting to move, and/or has little or no root penetration, is excluded.

Give depth to R layer or strongly cemented pan in metres.

BOUNDARIES BETWEEN HORIZONS/LAYERS

The nature and dimensions of the boundaries between horizons are determined visually. Accurate determination of boundaries is difficult in augered profiles. The thickness of the horizon and its adjoining horizons should be considered when determining horizon boundaries, particularly in the context of gradual and diffuse boundaries. Boundary shape is determined relative to the general (lateral) plane of the horizon boundary, which may or may not be horizontal. Examples are illustrated in Figure 23.

Boundary distinctness

		Width of boundary
S	*Sharp*	<5 mm
A	*Abrupt*	5–20 mm
C	*Clear*	20–50 mm
G	*Gradual*	50–100 mm
D	*Diffuse*	>100 mm

Boundary shape

S	*Smooth*	Horizon boundary is, or is almost, a plane surface.
W	*Wavy*	Horizon boundary contains undulations in the lateral plane with depressions wider than they are deep (i.e. a depth to width ratio ≤1).

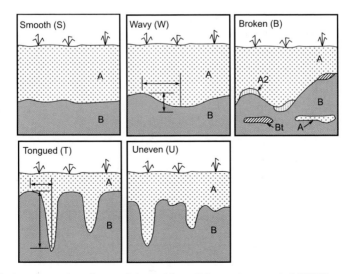

Figure 23 Horizon boundary shapes. Adapted from Schoeneberger *et al.* (2012).

T	*Tongued*[24]	Horizon boundary contains depressions in the lateral plane deeper than they are wide (i.e. a depth to width ratio >1).
U	*Uneven*	Horizon boundary contains undulations in the lateral plane that are inconsistent in both lateral and perpendicular dimensions.
B	*Broken*	Horizon boundary is discontinuous.

COLOUR

Record colours by comparing soils with colour charts using the Munsell Colour system – for example, the Munsell Soil Color Charts or the Revised Standard Soil Color Charts (Oyama and Takehara 1970). *Hue, value* and *chroma* of the matrix colour are recorded – for example, *10YR4/2*. Record the colour book used.

Colour book for colour description

| M | *Munsell* |
| J | *Revised Standard Color Charts* |

Soil colours rarely match the chart colour chips perfectly. They should be matched to the chip closest in colour, or the nearest whole number in chroma where chips are not provided (e.g. chromas 5 and 7).

The soil colour is measured on the surface of a freshly broken aggregate of moist soil under conditions of natural light. Moisten the soil if it is dry. Record the colour when the visible moisture film disappears from the surface of the moistened broken aggregate. The aggregate should be held as close as possible to the colour chips. Take care not to smear the broken surface, as this can give an incorrect recording of the colour of the soil matrix.

Dry colours may also be recorded and are required for the determination of a bleached horizon. Record the moisture status in conjunction with any colour.

Moisture status for colour description

| M | *Moist* |
| D | *Dry* |

MOTTLES AND OTHER COLOUR PATTERNS

Mottles are spots, blotches or streaks of subdominant colours different from the matrix colour and also different from the colour of the ped surface. Colour patterns, related to segregations of pedogenic origin (including root tracings), are not considered to be mottles and are recorded elsewhere (see page 166). Colour patterns due to biological or mechanical mixing and inclusions of weathered substrate material are described separately.

---

24 Tongued is encompassed within 'Irregular' in FAO (2006) and Schoeneberger *et al.* (2012).

Type

M	Mottles (individual)
R	Reticulate mottle patterns (strongly developed reddish, yellowish and greyish or white net-like patterns)
X	Colour patterns due to biological mixing of soil material from other horizons (e.g. worm casts)
Y	Colour patterns due to mechanical mixing of soil material from other horizons (e.g. inclusions of B horizon material in Ap horizons)
Z	Colour patterns due to inclusions of weathered substrate material

Abundance

The percentage is estimated by eye using the chart in Figure 18 for comparison (see page 83).

0	*No mottles or other colour patterns*	0
1	*Very few*	<2%
2	*Few*	2–10%
3	*Common*	10–20%
4	*Many*	20–50%

Size

Measure size along the greatest dimension, except in streaks or linear forms where width is measured.

1	*Fine*	<5 mm
2	*Medium*	5–15 mm
3	*Coarse*	15–30 mm
4	*Very coarse*	>30 mm

Contrast

F	*Faint*	Indistinct; evident only on close examination.
D	*Distinct*	Readily evident although not striking.
P	*Prominent*	Striking and conspicuous.

Colour

This should be described in terms of Munsell or Revised Standard Soil colours, but the following abbreviated forms can also be used:

R	*Red*
O	*Orange*
B	*Brown*

Y	Yellow	
G	Grey	
D	Dark	Values 3 or less and chromas 2 or less for all hues.
L	Gley	Gley charts only.
P	Pale	Values 7 or more and chromas 2 or less for all hues.

Distinctness of boundaries

S	Sharp	Knife-edge boundary between colours.
C	Clear	Colour transition over less than 2 mm.
D	Diffuse	Colour transition over 2 mm or more.

PARTICLE SIZE AND FIELD TEXTURE

The proportions and sizes of mineral matter that make up the soil matrix are represented in two main ways: field texture (a field measurement) and particle size (a laboratory measurement). The two measures are somewhat related and use some overlapping terms, but are discretely different methods resulting in differing observations of the soil material – one being qualitative and the other quantitative. There is only an approximate relationship between field texture and particle size distribution (see Marshall 1947), as factors other than clay, silt and sand content influence field texture.

In Australia, field texture classes or field texture grades (Northcote 1979) are based on field determination of texture and **not** on laboratory determinations of particle size. Both field texture and particle size are determined on the mineral size fraction finer than 2 mm. This is commonly referred to as the *fine-earth fraction*.

Particle size classes

The particle size classes used in laboratory analysis of soil materials differ between some disciplines (e.g. soil science *vs* engineering) and between countries (Figure 24). Historically, laboratory methods have been devised to yield results matching pre-determined particle size classes. Laser diffraction methods now in use yield a full particle size distribution (PSD). While they are not used in the determination of field texture, particle size classes in use are provided below as reference.

Texture triangles have long been in use to diagrammatically represent texture classes derived from particle size analysis data. The texture triangle adopted in Australia (Figure 25) is that of Marshall (1947), which is based on international size classes. Note that other texture triangles will potentially yield differences in particle size classes determined from any given data. The only difference between the recommended and international particle size classes in Figure 24 is within the sand class (fine sand/coarse sand boundary), which does not affect the use of Figure 25 with particle size data derived using the recommended classes. Marshall (1947) recognised the deficiencies of the texture triangle, particularly in relation to representation of differing sand size classes and proposed a second diagram for this purpose (Figure 26).

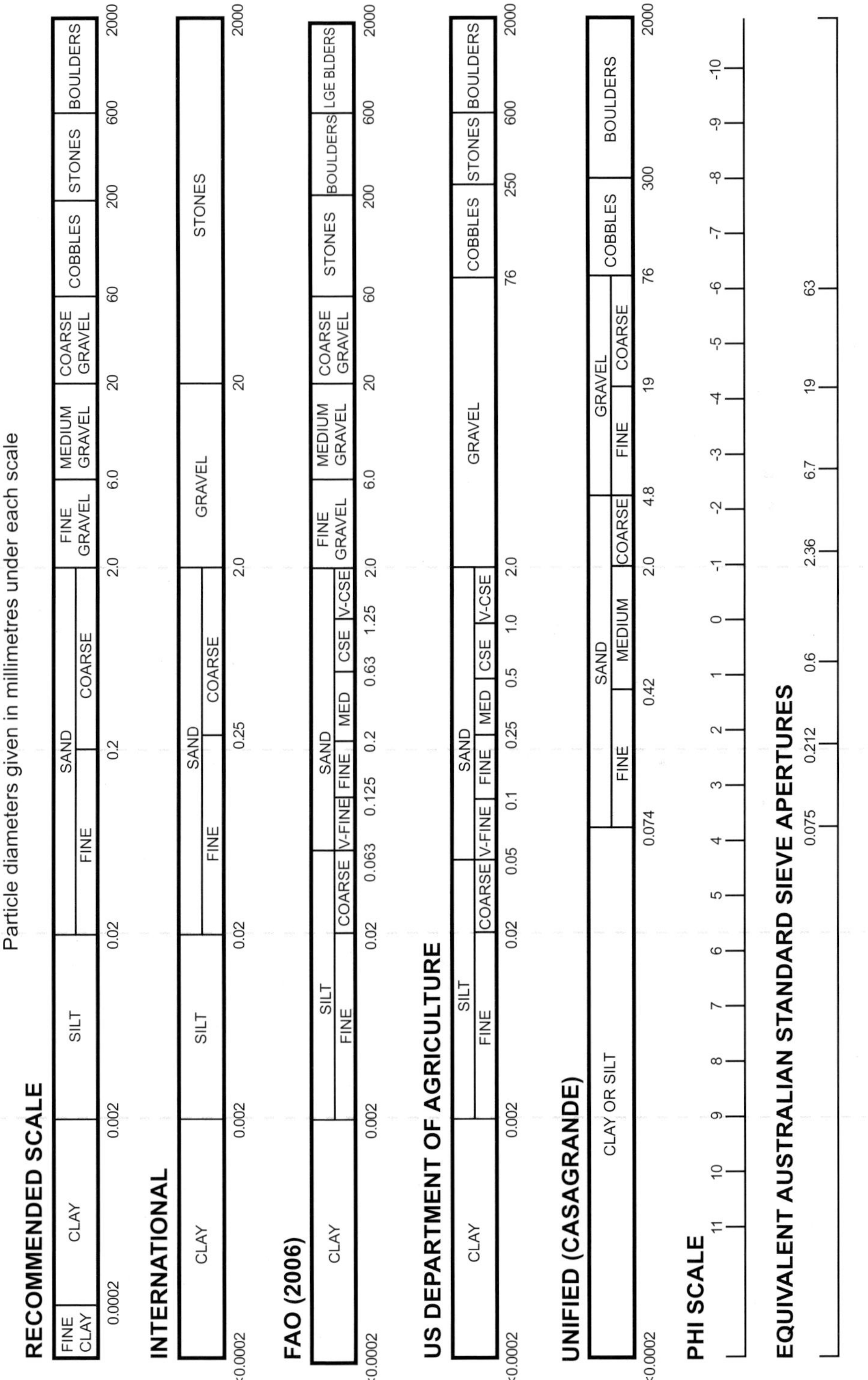

Particle diameters given in millimetres under each scale

Figure 24 Size fractions in several major classification systems.

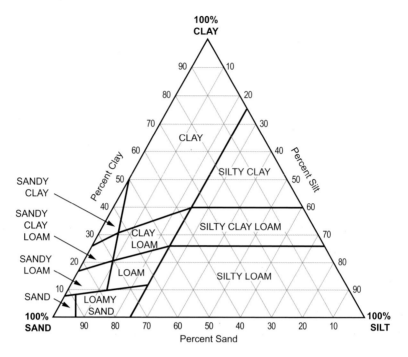

Figure 25 Triangular texture diagram based on international particle size fractions (Marshall 1947).

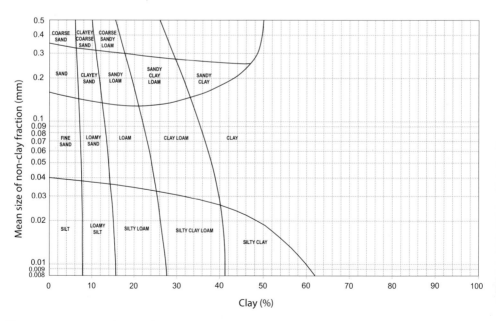

Figure 26 Rectangular texture diagram based on international particle size fractions (Marshall 1947).

Terminology

Prior to description of field textures, it is necessary to clarify terminology in use in particle size classes and field texture classes. The recommended classes in Figure 25 contain three base terms: *clay*, *silt* and *sand*. Note that while these are size classes, they are also recognisable entities. Clay, in the context of soils, implies secondary clay minerals. A range of secondary clay minerals are found in soils, and there is a well-established basis that in conventional particle size analysis methods these are the only components that occur at sizes <0.002 mm.

The sand size fraction will primarily be comprised of SiO_4, but may also contain other primary minerals, such as feldspar and mica, particularly in the coarse sand fraction. In Figure 24, sand is divided into fine and coarse fractions at 0.2 mm. However, in the determination of field texture, fine, medium and coarse sand are recognised. Fine sand and coarse sand as identified in the laboratory do not equate directly to that determined in the field, other than at the outer bounds of the sand size class.

In the field, the silt size fraction (0.02–0.2 mm) can be comprised of aggregated secondary clay minerals or very fine primary rock minerals. Silt, as a material, has varying definitions, from the layman to the geologist, which causes considerable confusion from a terminology perspective. In the pedological domain, silt (the material) is primarily derived from glacial action on rocks and is generally regarded as inert. It has recognisable field characteristics: pale colour, talcum-powder like behaviour (clings to skin), blows out of dry soil easily, low wet strength (quickly changes to 'wet slop' with the addition of water). As silt can suspend in water, its field behaviour in terms of erosion and slaking/dispersion tests can often appear to mirror dispersive clay soils. High silt fraction materials often demonstrate dilatancy (shear-thickening – their apparent viscosity increases with increasing shear rate). Accumulations of wind-blown silt are referred to as *loess*. In the laboratory, treatment methods used in particle size analysis ensure that aggregated clay is not a feature in the silt size fraction.

Field texture of mineral soils

Soil texture is determined by the size distribution of mineral particles finer than 2 mm, that is, only material that will pass a 2 mm sieve should be used in determination of field texture. Organic soils require a different approach to mineral soils and are discussed on page 145.

The following description of the determination of field texture of mineral soils is adapted from Northcote (1979). Field texture is a measure of the behaviour of a small handful of soil when moistened and kneaded into a ball and then pressed out between thumb and forefinger.

Take a sample of soil sufficient to fit comfortably into the palm of the hand. Moisten the soil with water, a little at a time, and knead until the ball of soil, so formed, just fails to stick to the fingers. Add more soil or water to attain this condition, known as the *sticky point*, which approximates field capacity for that soil. Continue kneading and moistening until there is no apparent change in the soil ball, usually a working time of 1–2 minutes. The soil ball, or bolus, is now ready for shearing manipulation, *but the behaviour of the soil during bolus formation is also indicative of its field texture. The behaviour of the bolus and of the ribbon produced by shearing (pressing out) between thumb and forefinger characterises the field texture. Do not assess field texture grade solely on the length of ribbon.*

The recommended field texture grades as characterised by the behaviour of the moist bolus are set out below. Note that the table does not include all possible field textures, when consideration is given to **modifiers** (sand size) and **qualifiers**. The approximate percentage content of clay (particles less than 0.002 mm in diameter) and silt (particles between 0.02 and 0.002 mm in diameter) are given as a guide. *These percentages must not be used to determine a field texture, that is, do not use them to convert a laboratory particle size value to a field texture grade. Similarly, do not adjust a field texture grade when laboratory particle size data become available.*

The three core terms in field textures are *sand*, *loam* and *clay*. The terms *sandy*, *loamy* and *clay(ey)* are also used to create texture classes that lie between core classes. Additional terms are appended as modifiers of sand size, or to identify the presence of silt; and as qualifiers (see page 143).

Field texture grade

		Behaviour of moist bolus	*Approximate clay content (%)*
S	*Sand*	Coherence nil to very slight, cannot be moulded; sand grains of medium size; single sand grains adhere to fingers.	Commonly less than 5%
LS	*Loamy sand*	Slight coherence; sand grains of medium size; can be sheared between thumb and forefinger to give minimal ribbon of about 5 mm.	~5%
CS	*Clayey sand*	Slight coherence; sand grains of medium size; sticky when wet; many sand grains stick to fingers; will form minimal ribbon of 5–15 mm; discolours fingers with clay stain.	5–10%
SL	*Sandy loam*	Bolus coherent but very sandy to touch; will form ribbon of 15–25 mm; dominant sand grains are of medium size and are readily visible.	10–20%
L	*Loam*	Bolus coherent and rather spongy; smooth feel when manipulated but with no obvious sandiness or 'silkiness'; may be somewhat greasy to the touch if much organic matter present; will form ribbon of about 25 mm.	~25%
ZL	*Silty loam*	Coherent bolus; very smooth to often silky when manipulated; will form ribbon of about 25 mm.	~25% and with silt 25% or more
SCL	*Sandy clay loam*	Strongly coherent bolus, sandy to touch; medium-size sand grains visible in finer matrix; will form ribbon of 25–40 mm.	20–30%
CL	*Clay loam*	Coherent plastic bolus, smooth to manipulate; will form ribbon of 40–50 mm.	30–35%
CLS	*Clay loam, sandy*	Coherent plastic bolus; medium-size sand grains visible in finer matrix; will form ribbon of 40–50 mm.	30–35%
ZCL	*Silty clay loam*	Coherent smooth bolus, plastic and often silky to the touch; will form ribbon of 40–50 mm.	30–35% and with silt 25% or more

LC	*Light clay*	Plastic bolus; smooth to touch; slight resistance to shearing between thumb and forefinger; will form ribbon of 50–75 mm.	35–40%
LMC	*Light medium clay*	Plastic bolus; smooth to touch; slight to moderate resistance to ribboning shear; will form ribbon of about 75 mm.	40–45%
MC	*Medium clay*	Smooth plastic bolus; handles like plasticine and can be moulded into rods without fracture; has moderate resistance to ribboning shear; will form ribbon of 75 mm or more.	45–55%
MHC	*Medium heavy clay*	Smooth plastic bolus; handles like plasticine; can be moulded into rods without fracture; has moderate to firm resistance to ribboning shear; will form ribbon of 75 mm or more.	50% or more
HC	*Heavy clay*	Smooth plastic bolus; handles like stiff plasticine; can be moulded into rods without fracture; has firm resistance to ribboning shear; will form ribbon of 75 mm or more.	50% or more

All of the above field texture grades in which sand is recorded – for example, LS, SL, SCL – are defined as having medium-sized sand. Coarse or fine sand grades can be given, as below:

K	*Coarse sandy*	Coarse sand is obviously coarse to touch. Sand grains are very readily seen with the naked eye.
F	*Fine sandy*	Fine sand can be felt and often heard when manipulated. Often sticks to fingers. Sand grains are clearly evident under a ×10 hand lens.

Record **K** or **F** immediately preceding S in the texture codes – for example, LKS, KSL, CLFS, FSL. This approach also applies to the use of the sand modifiers in the clay texture range. Either **KS, S** or **FS** may be used, as may **Z** (silty) – for example, KSLC (coarse sandy light clay), SMHC (sandy medium heavy clay), ZLMC (silty light medium clay).

Field texture qualification

The non-clay field texture grades (clay loams and coarser) may be qualified according to whether they are at, or near, the light (lower clay content) or heavy (higher clay content) end of the range for that particular field texture grade. Note that codes on the preceding page already provide for light and heavy qualifiers in the clay textures, hence – and + are not required in the clay texture range.

–	*Light*
+	*Heavy*

It is strongly recommended that this option only be used where considered essential. A considerable skill level is required to utilise these qualifiers consistently and accurately. If too freely used, it can lead to excessive, unnecessary detail of doubtful usefulness.

If soils are appreciably organic, but do not qualify as an organic pseudo-texture (see page 145), they are qualified thus:

I	*Fibric*	Organic and fibrous; dark organic stain discolours fingers; greasy feel in clayey textures and coherence in sandy textures. Fibres (excluding living roots) or plant tissue remain visible to the naked eye.
H	*Hemic*	Organic and semi-fibrous; dark organic stain discolours fingers; greasy feel in clayey textures and coherence in sandy textures. Fibres (excluding living roots) or plant tissue are visible with a ×10 hand lens.
A	*Sapric*	Organic and non-fibrous; dark organic stain discolours fingers; greasy feel in clayey textures and coherence in sandy textures. Fibres (excluding living roots) or plant tissue remains are *not* visible to the naked eye and little or none visible with a ×10 hand lens.

All qualifier codes are given *after* the field texture, although the plain English description places the term first – for example, SCL- (light sandy clay loam), SL+ (heavy sandy loam) and SLI (fibric sandy loam).

Soil properties affecting the determination of field texture grade

A number of soil properties affect field texture:

- *Clay* (particles less than 0.002 mm in diameter) confers cohesion, stickiness and plasticity to the bolus and increases its resistance to deformation.
- *The type of clay mineral* influences the tractability of the bolus. Montmorillonitic clays tend to make the bolus resist deformation and therefore it can be stiff to ribbon. Thus a long ribbon may suggest a finer (more clayey) field texture than the percentage clay content would indicate. By contrast, kaolinitic clays make the field texture appear less clayey than the percentage clay content would indicate, as they tend to produce a short thin ribbon from the bolus.
- *Silt* (particles between 0.002 and 0.02 mm in diameter) often confers a silky smoothness on field textures, as it fills in the particle size range between sand (particles more than 0.02 mm in diameter) and clay. The amount of silt required to give a silty 'feel', and thus invoke a silt qualifier, is likely to increase with increasing clay content.
- *Organic matter* confers cohesion to sandy field textures and a greasiness to clayey field textures; it tends to produce a short thick ribbon from the bolus. Some soils containing about 40–50% clay-sized particles and sufficient organic matter (over 20%) will behave as clay loams and light clays instead of medium or heavy clays. Large amounts of organic matter in dry soils may resist wetting and make bolus preparation difficult.
- When present in significant amounts, *oxides* – chiefly those of iron and aluminium – may require extra water for the soil to form the bolus. This may initially shear readily to produce a short ribbon, indicating a less clayey field texture than the clay content suggests. Such soil materials are *subplastic*.
- *Calcium and magnesium carbonates* in the fine earth fraction (particles less than 2 mm in diameter) will usually impart a porridge-like consistency to the bolus. They tend to increase the apparent clayeyness of sandy and loamy field textures such that amounts of 10–30%

calcium carbonate cause the field texture to increase about one grade above that obtained when the carbonates are removed from the fine earth fraction. Carbonates may also make clay field textures appear less clayey by shortening the ribbon produced from the bolus.

- *Cation composition.* In general, calcium-dominant clays accept water readily and are easy to knead and smooth to field texture. Sodium- and magnesium-dominant clays, on the other hand, are often difficult to wet and knead, producing a slimy, tough bolus, resistant to shearing and often appearing to have a more clayey field texture than would be indicated by the actual clay content.
- *Strong, fine-structural aggregation* (such as found in Ferrosols) will tend to cause an underestimate of clay content (subplastic properties), due to the incomplete breakdown of the structural units during bolus preparation. Longer and more vigorous kneading is necessary to produce a homogeneous bolus.

The above properties occur in soils to differing degrees and specific allowance cannot be made for them. Field texture must remain a subjective but reproducible measure of the behaviour of a handful of soil moistened and kneaded into an adequately prepared bolus and subjected to shearing manipulation between thumb and forefinger. However, this method does provide a very useful assessment of the physical behaviour of soil in the field.

Organic soil[25]

Strictly speaking, organic soil does not have a textural name, as soil texture is determined by the size of mineral particles finer than 2 mm (page 141), using criteria of the bolus such as feel and ribbon. From a practical perspective though, organic soil can have a pseudo-texture, related to the plant materials from which they formed, and the degree of decomposition, exposure and drying. Methods such as the von Post test (FAO 1988) may also be used to assist in characterisation of organic materials.

The following names may be used to characterise materials that on field examination are considered to be clearly dominated by organic matter. Use of these terms has a relationship to horizon designation (e.g. IP, HP and AP would invoke horizon designations of Pα, Pγ and Pθ respectively).

The following is adapted from Soil Survey Staff (1975) and Avery (1980):

IP	*Fibric organic soil*	Fibrous organic material – undecomposed or weakly decomposed organic material. Plant remains are distinct and readily identifiable.
HP	*Hemic organic soil*	Semi-fibrous organic material – moderately to well decomposed organic material. Plant remains vary from most being difficult to identify to most being unidentifiable. It is intermediate in degree of decomposition between the less decomposed fibric organic material and the more decomposed, sapric organic material.
AP	*Sapric organic soil*	Humified organic material – strongly to completely decomposed organic material. Plant remains vary from few being identifiable to completely amorphous.

---

25 Note the discussion regarding the quantitative definition of peat on page 126 and refer to Appendix 2.

If the soils are appreciably organic but have a sub-dominant mineral component, they may be qualified with a mineral texture (modifiers of coarse and fine for sand can also be used), in the same way that organic qualifiers and modifiers are applied to mineral textures.

SP	*Sandy organic soil*	Organic material in which the bolus is sandy to touch (but still dominated by organic material).
LP	*Loamy organic soil*	Organic material in which the bolus has obvious mineral particle content but no obvious sandiness to touch and is smooth, non-sticky when wet, and weakly coherent (but still dominated by organic material).
ZP	*Silty organic soil*	Organic material in which the bolus has obvious mineral particle content but no obvious sandiness to touch and is smooth, silky and weakly to moderately coherent (but still dominated by organic material). This may be difficult to distinguish from the nature of sapric organic material.
CP	*Clayey organic soil*	Organic material in which the bolus has obvious fine mineral particle content, is sticky when wet, and coherent (but still dominated by organic material).

On drying, peat may change irreversibly.

GP	*Granular peat*	Peat that has dried irreversibly to fine granules through exposure and drying and/or cultivation. Granules are approximately 1–2 mm in diameter and have a granular or subangular blocky structure.

Horizons with these pseudo-textures will still qualify as P horizons.

STRUCTURE

Soil structure refers to the distinctness, size and shape of peds. A ped is an individual *natural* soil aggregate consisting of a cluster of primary particles. Peds are separated from adjoining peds by surfaces of weakness that are recognisable as natural voids or by the occurrence of cutans (Brewer 1960).

Soil structure can only be described reliably in a relatively fresh vertical exposure or relatively undisturbed soil core, not from an auger boring. Vertical exposures that have been exposed for a long time (road cuttings, gullies) are unsuitable for the determination of structure (unless cut back to a clean face), as the structure may alter due to daily or seasonal changes in moisture and temperature. In any instance, moisture content may influence the degree to which structural units are evident, and in some cases will influence the size of visible peds. Narrow soil cores may limit observation of structural units larger in dimension than the core.

Compound pedality

Compound pedality occurs where large peds part along natural planes of weakness to form smaller peds, which may again part to smaller peds, and so on, to the smallest or primary peds. Compound pedality is common in structured soils, thus it is appropriate to describe more than one size.

Primary peds are the simplest peds occurring in soil material; they cannot be divided into smaller peds, but may be packed together to form compound peds of a higher level of organisation (Brewer 1964). The order of peds and relationship of one to the other is important and may be described as the larger peds parting to the smaller and further where necessary – for example, 'strong 50–100 mm columnar, parting to moderate 20–50 mm prismatic, parting to moderate 10–20 mm angular blocky'. The word 'parting' and not 'breaking' is used. The term 'breaking' is used when soil is fractured along planes other than natural planes of weakness. Always consider the largest structural units first and whether or not compound pedality exists.

Describe compound pedality in order as follows:

1   *Largest peds* (in the type of soil observation described), parting to

2   *Next size peds*, parting to

3   *Next size peds*, ... and further, if required, to the primary ped.

Grade of pedality

Grade of pedality is the degree of development and distinctness of peds. In virtually all material that has structure, the surface of individual peds will differ in some way from the interior of peds. The degree of development expresses the relative difference between the strength of cohesion within peds and the strength of adhesion between adjacent peds. Determination of grade of structure in the field depends on the proportion of peds that hold together as entire peds when displaced, and also on the ease with which the soil separates into discrete peds.

Grade of pedality varies with the soil water status. It is important to record the soil water status of the described profile (page 161), and it is desirable that the grade of pedality be described at the soil water status most common for the horizon.

Apedal soils have no observable peds and are divided into:

G   *Single grain*   Loose, incoherent[26] mass of individual particles. When displaced, soil separates into ultimate particles.

V   *Massive*   Coherent.[27] When displaced, soil separates into fragments, which may be crushed to ultimate particles.

Pedal soils have observable peds and are divided into:

W   *Weak*   Peds are indistinct and barely observable in undisplaced soil. When displaced, up to one-third of the soil material consists of peds, the remainder consisting of variable amounts of fragments and ultimate particles.

---

26   Incoherent means that less than two-thirds of the soil material will remain united at the given moisture state without very small force (force 1, see 'Consistence' on page 162) having been applied.

27   Coherent means that two-thirds or more of the soil material will remain united at the given moisture state unless force is applied.

| M | *Moderate* | Peds are well-formed and evident but not distinct in undisplaced soil. When displaced, more than one-third (but less than two-thirds) of the soil material consists of entire peds, the remainder consisting of broken peds, fragments and ultimate particles. |
| S | *Strong* | Peds are quite distinct in undisplaced soil. When displaced, more than two-thirds of the soil material consists of entire peds. |

Size of peds

The *average least dimension of peds* is used to determine the class interval. Use Figure 27 on pages 149–154 as a guide.

The least dimension is the *vertical* dimension for platy structure; the *horizontal* dimension for prismatic, columnar, blocky and polyhedral peds; the *maximum separation* (thickness) of convex faces for lenticular peds; and the *diameter* for granular peds. Determination of ped size should be cognisant of the method of exposure of the soil profile.

1	*<2 mm*
2	*2–5 mm*
3	*5–10 mm*
4	*10–20 mm*
5	*20–50 mm*
6	*50–100 mm*
7	*100–200 mm*
8	*200–500 mm*
9	*>500 mm*

Type of pedality

The types of structure are described below and illustrated in Figure 28.

CA	*Cast*	Faunal casts are soil aggregates created directly by living creatures, unlike other forms of aggregates, in which formation is dominated by soil physical and chemical processes. Casts are formed from, or are deposited in/on, the soil profile and include:
		• excreta of soil fauna which may be discrete particles – for example, insect faeces or the dense, coherent, globular forms of earthworm excreta. They are generally spherical or ovate in shape and have a strong conchoidal fracture.
		• soil masticated with salivary secretions into globular forms, for example, by ants, crickets and wasps.
GR	*Granular*	Spheroidal with limited accommodation to the faces of surrounding peds.
PL	*Platy*	Soil particles arranged around a horizontal plane and bounded by relatively flat horizontal faces with much accommodation to the faces of surrounding peds.

Platy Peds

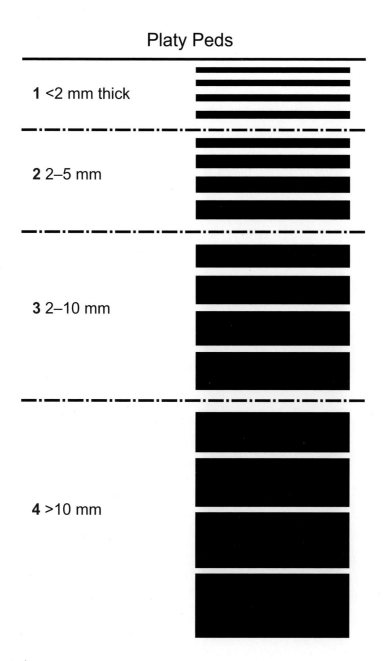

1 <2 mm thick

2 2–5 mm

3 2–10 mm

4 >10 mm

Figure 27 Ped size.

Prismatic and Columnar Peds

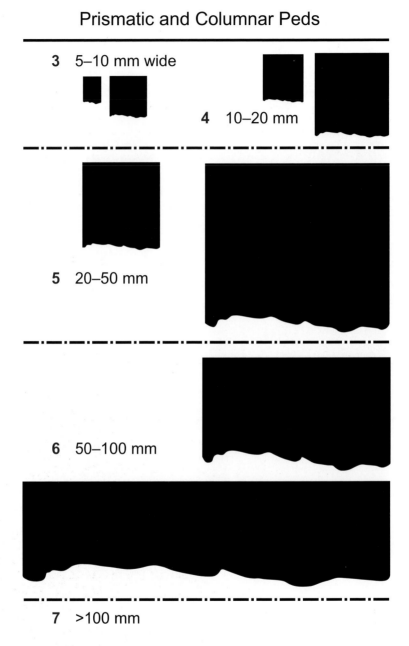

3 5–10 mm wide

4 10–20 mm

5 20–50 mm

6 50–100 mm

7 >100 mm

Figure 27 (continued) Ped size.

Angular and Subangular Blocky Peds

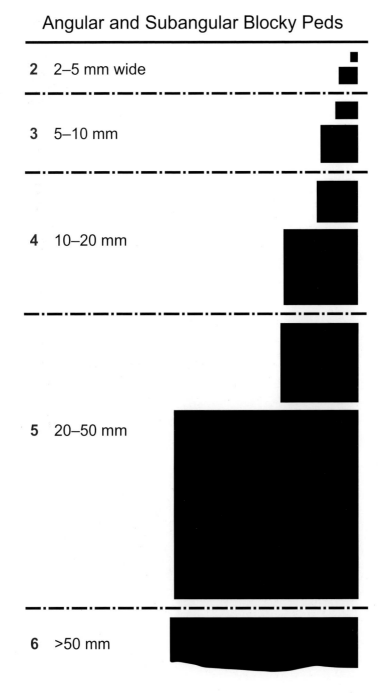

2 2–5 mm wide

3 5–10 mm

4 10–20 mm

5 20–50 mm

6 >50 mm

Figure 27 (continued) Ped size.

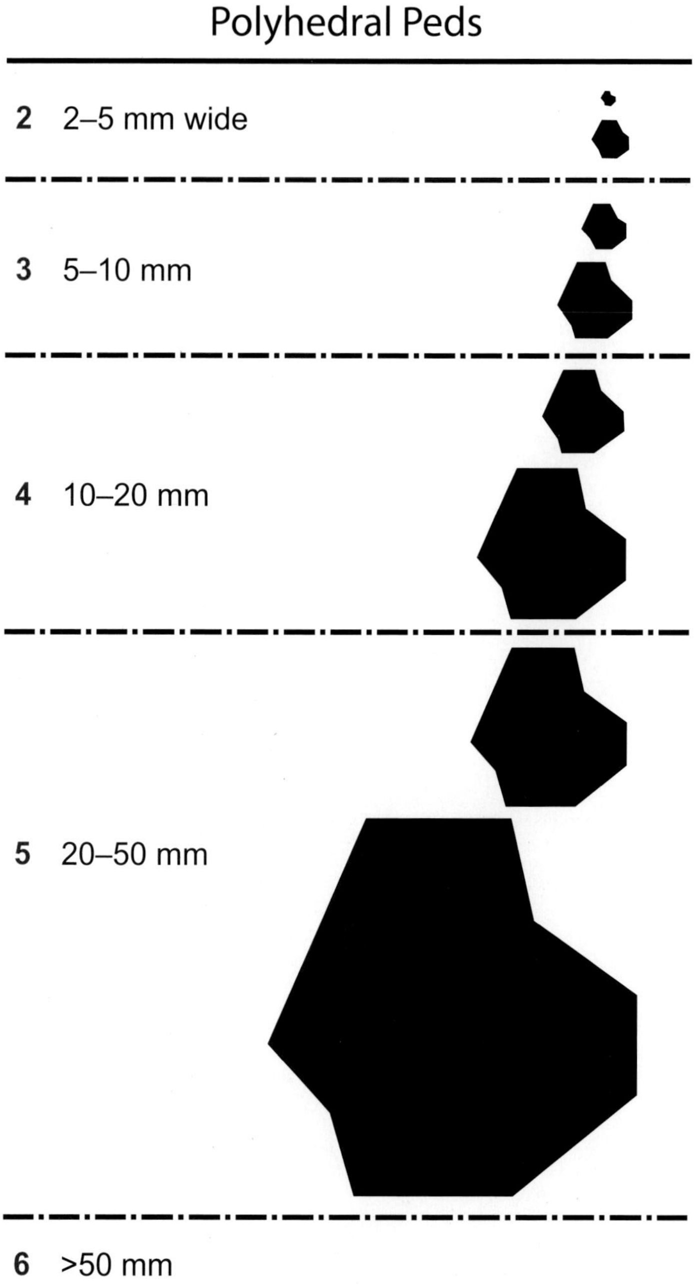

Polyhedral Peds

2 2–5 mm wide

3 5–10 mm

4 10–20 mm

5 20–50 mm

6 >50 mm

Figure 27 (continued) Ped size.

Lenticular Peds

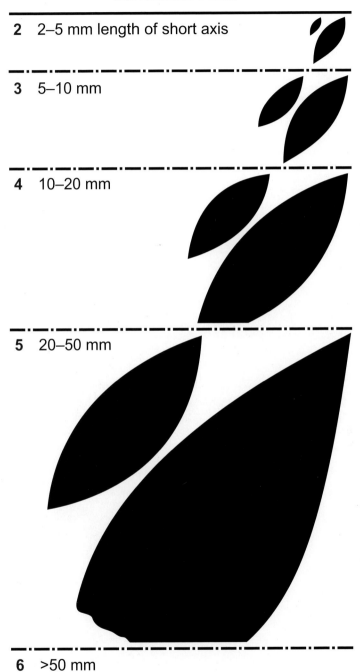

2 2–5 mm length of short axis

3 5–10 mm

4 10–20 mm

5 20–50 mm

6 >50 mm

Figure 27 (continued) Ped size.

Granular Peds

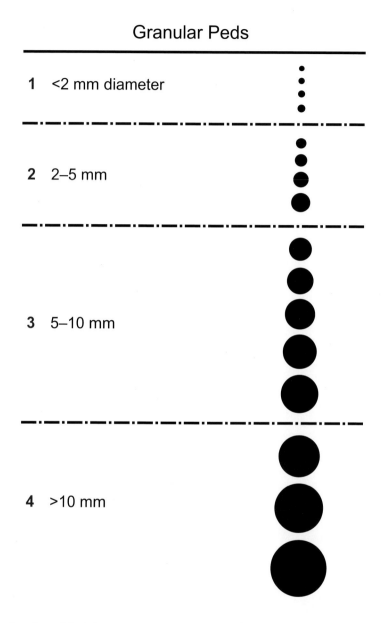

1 <2 mm diameter

2 2–5 mm

3 5–10 mm

4 >10 mm

Figure 27 (continued) Ped size.

PO   *Polyhedral*   Soil particles arranged around a point and bounded by more than six relatively flat, unequal, dissimilar faces. Re-entrant angles between adjoining faces are a feature. There is usually much accommodation of ped faces to the faces of surrounding peds. Most vertices are angular.

SB   *Subangular blocky*   Similar to *angular blocky* except peds are bounded by flat and rounded faces with limited accommodation to the faces of surrounding peds. Many vertices are rounded.

AB   *Angular blocky*   Soil particles arranged around a point and bounded by six relatively flat, roughly equal faces. Re-entrant angles between adjoining faces are few or absent. There is usually much accommodation of ped faces to the faces of surrounding peds. Most vertices between adjoining faces are angular.

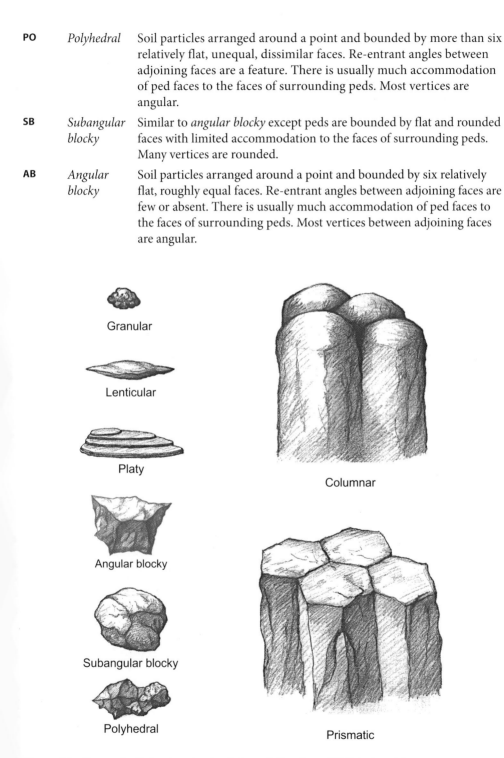

Figure 28   Types of soil structure (cast not shown). Courtesy of N. Schoknecht.

LE	*Lenticular*	Soil particles arranged around an elliptical or circular plane and bounded by curved faces with much accommodation to the faces of surrounding peds. Most vertices are angular and acute. This type also includes wedge-shaped peds, as they are a related shape and the two shapes are often found intermingled.
PR	*Prismatic*	Soil particles arranged around a vertical axis and bounded by well-defined, relatively flat faces with much accommodation to the faces of surrounding peds. Vertices between adjoining faces are usually angular.
CO	*Columnar*	As for *prismatic* but with domed tops.

Clods

Cultivated horizons (Ap horizons) often consist of artificial aggregates (clods). Tillage of soils above the plastic limit can lead directly to the creation of artificial aggregates, or they may be formed from fracturing of previously compacted layers. The distinction between artificial aggregates and peds can be difficult, but in clay soils, artificial aggregates usually possess a characteristic dull surface and conchoidal fracture planes. In cultivated horizons, where the pedologist is confident the aggregates are natural peds, they should be recorded as such. If the pedologist is doubtful, or the aggregates are obviously artificial, they should be recorded as clods. Presence of clods in a ploughed horizon would trigger the use of the p suffix (see page 132).

CL	*Large clod*	Artificial aggregate with diameter >100 mm.
FR	*Small clod*	Artificial aggregate with diameter <100 mm.

FABRIC

The following description is adapted from Northcote (1979). Fabric describes the appearance of the soil material (under a ×10 hand lens). Differences in fabric are associated with the presence or absence of peds, the lustre or lack of lustre of the ped surfaces, and the presence, size and arrangement of pores (voids) in the soil mass. The descriptions given below apply primarily to B horizons.

Organic fabric

The soil is dominated by organic materials. Peds are generally absent and the fabric varies with the nature and degree of decomposition of the organic materials (see P horizons and organic soil). It may be coherent to incoherent and may have visible layering.

Earthy (or porous) fabric

The mineral soil material is coherent and characterised by the presence of pores (voids) and few, if any, peds. Ultimate soil particles (e.g. sand grains) are coated with oxides and/or clays and are arranged (clumped) around the pores.

Sandy fabric

The mineral soil material is coherent, with few, if any, peds. The closely packed sand grains provide the characteristic appearance of the soil mass.

Rough-ped fabric

Peds are evident, and characteristically more than 50% of the peds are *rough-faced* – they have relatively porous surfaces. Rough-faced peds generally have less clearly defined faces than smooth-faced peds and the pedality of the soil may be questioned. However, if the soil mass is pressed gently, the characteristic size and shape of the soil aggregates will confirm its pedality. Granular peds with common or many macropores are always rough-faced, but this condition varies in other ped forms.

Smooth-ped fabric

Peds are evident, and characteristically more than 50% of them are dense and *smooth-faced*, although the degree of lustre may vary.

E	*Earthy*
G	*Sandy (grains prominent)*
R	*Rough-ped*
S	*Smooth-ped*

CUTANS

A cutan is a modification of the texture, structure or fabric within natural surfaces in soil materials; it arises from concentration of particular soil constituents or *in situ* modification of the plasma. Cutans comprise any of the component substances of the soil material (Brewer 1964).

Cutans may be observed in the field (a ×10 hand lens is usually necessary), but their nature is often difficult to determine unless a thin section is made. Hence the following simple classification.

Types of cutans

Z	*Zero or no cutans*	
U	*Unspecified*	Nature of cutans cannot be determined.
C	*Clay skins*	Coatings of clay often different in colour from the matrix of the ped. They are frequently difficult to distinguish from *stress cutans*, which are not true coatings.
M	*Mangans*	Coatings of manganese oxides or hydroxides. The material may have a glazed appearance and is very dark brown to black.
S	*Stress cutans*	*In situ* modifications of natural surfaces in soil materials due to differential forces such as shearing. They are not true coatings.
K	*Slickensides*	Stress cutans with smooth striations or grooves (typically associated with lenticular structure).
L	*Lamellae*	Thin, often discontinuous layers of clay-enriched material.
O	*Other cutans*	May be composed of iron oxides, organic matter, calcium carbonate or gypsum.

Abundance of cutans

0	*No cutans*	0
1	*Few*	<10% of ped faces or pore walls display cutans.
2	*Common*	10–50% of ped faces or pore walls display cutans.
3	*Many*	>50% of ped faces or pore walls display cutans.

Distinctness of cutans

This refers to the ease and certainty with which a cutan is identified. Distinctness relates to thickness and to the colour contrast with the adjacent material; it may change markedly with moisture content.

F	*Faint*	Evident only on close examination with ×10 magnification. Little contrast with adjacent material.
D	*Distinct*	Can be detected without magnification. Contrast with adjacent material is evident in colour, texture or other properties.
P	*Prominent*	Conspicuous without magnification when compared with a surface broken through the soil. Colour, texture or some other property contrasts sharply with properties of the adjacent material, or the feature is thick enough to be conspicuous.

VOIDS

This is a general term for pore spaces and other openings in soils, not occupied by solid mineral matter. The most important are *cracks* (planar voids) and *pores*, which are approximately circular in cross-section. The following types of voids have been adapted from FAO (2006).

Types of voids

C	*Channels*	Elongated voids of faunal or floral (plant root) origin, mostly tubular in shape and continuous, varying strongly in diameter. When wider than a few centimetres (burrow holes), they are more adequately described under microrelief.
I	*Interstitial*	Also known as textural voids. Predominantly irregular in shape and interconnected, and hard to quantify in the field. Controlled by the fabric, or arrangement, of the soil particles. Subdivision is possible into simple packing voids, which relate to the packing of sand particles, and compound packing voids, which result from the packing of non-accommodating peds.
P	*Planar*	Most planes (cracks) are extra-pedal voids, related to accommodating ped surfaces or cracking patterns. They are often not persistent and vary in size, shape and quantity depending on the moisture condition of the soil. Planar voids may be recorded by describing width and frequency.

V	*Vesicular*	Discontinuous, spherical or elliptical voids (chambers) of sedimentary origin or formed by compressed air (e.g. gas bubbles in slaking crusts after heavy rainfall).

Planar void (crack) dimension

		Width
1	*Fine*	<5 mm
2	*Medium*	5–10 mm
3	*Coarse*	10–20 mm
4	*Very coarse*	20–50 mm
5	*Extremely coarse*	>50 mm

Non-planar void (pore) dimension

Pores are divided into:

- micropores less than 0.075 mm diameter
- macropores greater than 0.075 mm diameter.

Figure 29 illustrates the size classes. Only macropores can be seen with the naked eye. All visible pores, holes, channels and tubes *within* peds, clods, fragments or apedal soil are recorded in the classes below.

Abundance of macropores

There are two groups:

Very fine and fine macropores (less than 2 mm diameter):

0	*No very fine or fine macropores*		
1	*Few*	<1 per 100 mm²	(10 mm × 10 mm)
2	*Common*	1–5 per 100 mm²	(10 mm × 10 mm)
3	*Many*	>5 per 100 mm²	(10 mm × 10 mm)

Medium and coarse macropores (greater than 2 mm diameter):

0	*No medium or coarse macropores*		
4	*Few*	<1 per 0.01 m²	(100 mm × 100 mm)
5	*Common*	1–5 per 0.01 m²	(100 mm × 100 mm)
6	*Many*	>5 per 0.01 m²	(100 mm × 100 mm)

Diameter of macropores

Use Figure 29 as a guide to average diameter.

1	*Very fine*	0.075–1 mm
2	*Fine*	1–2 mm

Size of Macropores

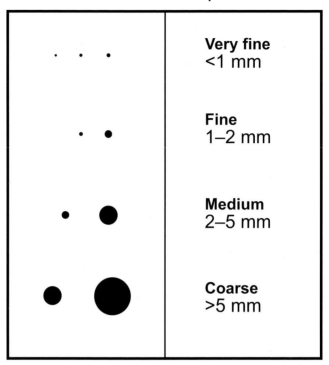

Very fine
<1 mm

Fine
1–2 mm

Medium
2–5 mm

Coarse
>5 mm

Figure 29   Size of macropores.

3	*Medium*	2–5 mm
4	*Coarse*	>5 mm

COARSE FRAGMENTS

Coarse fragments may occur throughout the profile. Their abundance, size, shape, lithology, strength, alteration and magnetic attributes are described in the same way as coarse fragments on the surface (see page 113); additionally, their distribution is described.

Coarse fragment distribution

U	*Undisturbed*	All fragments are remnants of the underlying bedrock and their orientation closely parallels that of the joint or bedding planes of the bedrock.
R	*Reoriented*	All fragments are remnants of the underlying bedrock but their orientation is not related to the joint or bedding planes of the bedrock.
S	*Stratified*	Fragments occur in bands, usually parallel with the soil surface (excluding those parallel with joint or bedding planes of the bedrock). They may include materials other than those from the underlying bedrock.

D   *Dispersed*   Fragments are scattered randomly throughout the soil and may be of mixed origin.

SOIL WATER STATUS

Describe soil water status of the soil at the time of description (see Table 31). It may also be relevant to note the weather conditions immediately prior to examination of the soil if these are known. For example, a soil may be wet because of local rain or from seepage.

The following guidelines may be used as a crude approximation of soil water status:

* *Dry* is below wilting point (crop lower limit). Material becomes darker or has lower colour value when moistened.
* *Moderately moist* is the drier half of the available moisture range.
* *Moist* is the wetter half of the available moisture range.
* *Wet* is at, or exceeding, field capacity (drained upper limit). Will wet and/or stick to fingers when moulded.

These guidelines may not apply with sodic 2:1 clays, as, for example, they may be moderately moist but below wilting point. Wet soils do not require water to be added in order to achieve a field texture.

CONSISTENCE

Consistence refers to the strength of cohesion and adhesion in soil. Strength will vary according to soil water status. Note that soil water status *must* be recorded with strength.

Strength

Strength of soil is the resistance to breaking or deformation. It may be measured quantitively in the field with instruments such as penetrometers (Hignett 2002). Strength is determined qualitatively by the force just sufficient to break or deform a 20 mm diameter piece of soil when a compressive shearing force is applied between thumb and forefinger. The 20 mm piece of soil may be a ped, part of a ped, a compound ped or a fragment of an apedal (massive) soil.

Table 31   Soil water status

		Behaviour of soils subjected to field test		
Soil water status		**Sands, sandy loams**	**Loams**	**Clay loams, clays**
D	*Dry*	Will flow through fingers or fragments will powder.	Will not ball when squeezed in hand. Fragments will powder.	Will not ball when squeezed in hand. Fragments will break to smaller fragments or peds.
T	*Moderately moist*	Appears dry. Ball will not hold together.	Forms crumbly ball on squeezing in hand.	Will ball. Will not ribbon.
M	*Moist*	Forms weak ball but breaks easily.	Will ball. Will not ribbon.	Will ball. Will ribbon easily.
W	*Wet*	Ball leaves wet outline on hand when squeezed, or is wetter.	Ball leaves wet outline on hand when squeezed, or is wetter. Sticky.	Ball leaves wet outline on hand when squeezed, or is wetter. Sticky.

Force[28]

0	*Loose*	No force required. Separate particles such as loose sands.
1	*Very weak*	Very small force, almost nil.
2	*Weak*	Small but significant force.
3	*Firm*	Moderate or firm force.
4	*Very firm*	Strong force but within power of thumb and forefinger.
5	*Strong*	Beyond power of thumb and forefinger. Crushes underfoot on hard, flat surface with small force.
6	*Very strong*	Crushes underfoot on hard, flat surface with full body weight applied slowly.
7	*Rigid*	Cannot be crushed underfoot by full body weight (average person) applied slowly.

Stickiness

Stickiness is determined on *wet* soil by pressing the wet sample between thumb and forefinger and then observing the adherence of the soil to the fingers.

0	*Non-sticky*	Little or no soil adheres.
1	*Slightly sticky*	Soil adheres to thumb and forefinger but is not stretched notably and comes off rather cleanly.
2	*Moderately sticky*	Soil adheres to thumb and forefinger and tends to stretch rather than pull free of fingers.
3	*Very sticky*	Soil adheres strongly to thumb and forefinger and stretches notably.

Type of plasticity

Plasticity is the ability to change shape and retain the new shape after the stress is removed. The type of plasticity refers to the degree to which the consistence and/or field texture properties of a soil suggest the amount of clay-sized particles it contains (Butler 1955). It may be identified by determining two field textures: one after an initial 1 to 2 minute working of the soil sample, and another after a prolonged 10 minute kneading. The change in field texture from the initial to the prolonged working of the soil sample indicates the type of plasticity.

Field texture change after 10 minute kneading

| S | *Superplastic* | Decreases one or more field texture groups of Northcote (1979). |
| N | *Normal plasticity* | Negligible change. |

---

28 'Forces 0 to 5 are equivalent to the following dry consistence classes in the 'USDA Soil Survey Manual' (Soil Survey Staff 1951):

0	*Loose*	3	*Hard*
1	*Soft*	4	*Very hard*
2	*Slightly hard*	5	*Extremely hard*

| U | *Subplastic* | Increases one to two field texture groups. |
| T | *Strongly subplastic* | Increases two or more field texture groups. |

Degree of plasticity

The degree of plasticity is determined at the soil moisture content used for field texturing, that is, just below sticky point. The soil is rolled between the palms of the hand and, if possible, 40 mm long rolls are formed. The rolls are dangled from the thumb and forefinger. Plasticity is determined on the following scale of behaviour of rolls of varying thickness (note: the degree of plasticity given below applies only to normal plasticity).

		Dimensions and behaviour of 40 mm rolls	
		Diameter	Behaviour
0	*Non-plastic*	6 mm	Will not form.
1	*Slightly plastic* 6 mm	4 mm	Will form but will not support its own weight.
		Will form and will support its own weight.	
2	*Moderately plastic* 4 mm	2 mm	Will form but will not support its own weight.
		Will form and will support its own weight.	
3	*Very plastic*	2 mm	Will form and will support its own weight.

PANS

A pan is an *indurated or cemented soil horizon or layer*. Description of a pan has an associated relationship with the use of suffixes such as *h, s, m* and *q*. A pan is typically a B horizon, but at times it may be difficult to discern if a pan is a paleo-feature.

Cementation of pan

Place a 30 mm diameter piece of the pan in water for 1 hour. If it softens or slakes, it is uncemented; if not, it is cemented. The degree of cementation is assessed on the following scale after 1 hour soaking in water.

0	*Not cemented*	Slakes or softens so that only very weak force is required to crush.
1	*Weakly cemented*	Small but significant force is required to crush. Within power of thumb and forefinger.
2	*Moderately cemented*	Beyond power of thumb and forefinger. Crushes underfoot on hard, flat surface with weight of average person (80 kg) applied slowly.
3	*Strongly cemented*	Cannot be crushed underfoot by weight of average person (80 kg) applied slowly. Can be broken by hammer.

| **4** | *Very strongly cemented* | Cannot be broken by hammer, or only with extreme difficulty. |

Type of pan

Z	*Zero or no pan*	
A	*Alcrete (bauxite)*	Indurated material rich in aluminium hydroxides. Commonly consists of cemented pisoliths and usually known as bauxite. May invoke use of the *cc* suffix on the horizon.
C	*Organic pan*	Horizon relatively high in organic matter but low in iron. It is relatively thick and weakly to strongly cemented by aluminium and usually becomes progressively more cemented with depth. It is usually relatively uniform in appearance laterally. It is commonly the B horizon of humus Podosols, where it is often known as coffee rock or sandrock. Typically invokes use of the *h* suffix on the horizon.
D	*Duripan*	Earthy pan so cemented by silica that dry fragments do not slake in water and are always brittle, even after prolonged wetting (described by Soil Survey Staff 1975). Duripans occur mostly in arid to semi-arid climates. Typically invokes use of the *qm* suffix on the horizon.
E	*Ferricrete*	Indurated material rich in hydrated oxides of iron (usually goethite and hematite) occurring as cemented nodules and/or concretions, or as massive sheets. This material has been commonly referred to in local usage around Australia as laterite, duricrust or ironstone. May invoke use of the *cc* suffix on the horizon.
F	*Fragipan*	Earthy pan, which is usually loamy. A dry fragment has moderate or weak brittleness and slakes in water. A wet fragment does not slake in water. Fragipans are more stable on exposure than overlying or underlying horizons (described by Soil Survey Staff 1975).
H	*Hardsetting pan*	Hard, apedal soil condition that forms on drying but softens on wetting and typically occurs in sands. When dry, the material is hard and is not disturbed or indented by pressure of forefinger. Hardsetting pans occur as discrete horizon(s) that are harder in the dry state and more stable on exposure than overlying and underlying layers.
I	*Thin ironpan*	Commonly thin (2–10 mm), black to dark reddish pan cemented by iron, or iron and manganese, or iron–organic matter complexes. Rarely 40 mm thick. Has wavy or convolute form and usually occurs as a single pan. This is a placic horizon (described by Soil Survey Staff 1975, (page 33)).
K	*Calcrete*	Any cemented, terrestrial carbonate accumulation that may vary significantly in morphology and degree of cementation. Also known as carbonate pan, calcareous pan, caliche, kunkar, secondary limestone, travertine. All show slight to strong effervescence with 1-molar HCl. May invoke use of the *kk* suffix on the horizon.

L	*Silcrete*	Strongly indurated siliceous material cemented by, and largely composed of, forms of silica, including quartz, chalcedony, opal and chert. Typically invokes use of the *qm* suffix on the horizon.
M	*Manganiferous pan*	Indurated material dominated by oxides of manganese. May invoke use of the *cc* suffix on the horizon.
N	*Densipan*	Earthy pan, which is very fine sandy (0.02–0.05 mm). Fragments, both wet and dry, slake in water. Densipans are less stable on exposure than overlying or underlying horizons (described by Smith *et al.* 1975).
R	*Red-brown hardpan*	Earthy pan, which is normally reddish-brown to red with dense yet porous appearance; it is very hard, has an irregular laminar cleavage and some vertical cracks, and varies from less than 0.3 m to over 30 m thick. Other variable features are bedded and unsorted sand and gravel lenses; wavy black veinings, probably manganiferous; and, less commonly, off-white veins of calcium carbonate (the presence of calcium carbonate is not common and the red-brown hardpan in which it occurs may be relatively brittle and finely laminar). The red-brown hardpan is usually present below the soil profile and is not a feature of any particular soil group. It has some similarity with other silica pans such as duripans of arid climates. In many instances, it is not known if the red-brown hardpan is a paleosol or a cemented sediment (Wright 1983). If thought to be the latter, it may be described as a substrate material (page 177). Typically invokes use of the *qm* suffix on the horizon.
T	*Ortstein*	Horizon strongly cemented by iron and organic matter. It has marked local variability in colour, both laterally and vertically. It may occur in the B horizon of podzols. Typically invokes use of the *hs* suffix on the horizon.
V	*Cultivation pan*	Subsurface soil horizon having higher bulk density, lower total porosity, and lower permeability to both air and water than horizons directly above and below as a result of cultivation practices (after Morse *et al.* 1987).
Y	*Gypsum pan*	Horizon strongly cemented by gypsum. May invoke use of the *yy* suffix on the horizon.
O	*Other pans*	Pans not listed above.

Continuity of pan

C	*Continuous*	Extends as a layer with little or no break across 1 m or more.
D	*Discontinuous*	Broken by cracks but original orientation of fragments is preserved.
B	*Broken*	Broken by cracks and fragments are disoriented.

Structure of pan

V	*Massive*	No recognisable structure.
S	*Vesicular*	Sponge-like structure having large pores, which may or may not be filled with softer material.
C	*Concretionary*	Spheroidal concretions cemented together.
N	*Nodular*	Nodules of irregular shape cemented together.
L	*Platy*	Plate-like units cemented together.
R	*Vermicular*	Worm-like structure and/or cavities.

SEGREGATIONS OF PEDOGENIC ORIGIN

This refers to discrete segregations that have accumulated in the soil because of the concentration of some constituent, usually by chemical or biological action. Segregations may be relict or formed *in situ* by current pedogenic processes (including biologically mediated reactions). Specific methods can be used to confirm the nature of some segregations in the field. These include: the use of hydrochloric acid (HCl) to confirm carbonate segregations and hydrogen peroxide (H_2O_2) to confirm manganiferous segregations (see pages 169–170).

Abundance of segregations
Use Figure 18 as a guide (see page 114).

0	*No segregations*	0
1	*Very few*	<2%
2	*Few*	2–10%
3	*Common*	10–20%
4	*Many*	20–50%
5	*Very many*	>50%

Nature of segregations

A	*Aluminous (aluminium)*
E	*Earthy (dominantly non-clayey)*
F	*Ferruginous (iron)*
G	*Ferruginous–organic (iron–organic matter)*
H	*Organic (humified, well-decomposed organic matter)*
K	*Calcareous (carbonate)*
L	*Argillaceous (clayey)*
M	*Manganiferous (manganese)*
N	*Ferromanganiferous (iron–manganese)*

O	Other
S	Sulphurous (sulfur), for example jarosite in acid sulphate soils
U	Unidentified
Y	Gypseous (gypsum)
Z	Saline (visible salt that is not gypsum)

Form of segregations

The form of a segregation encompasses aspects of the genesis, morphology and shape. In some cases, it may be appropriate to specifically record shape – for example, crystalline gypsum forms can adopt a variety of shapes, including rosettes, blades (planar) and needles. Some forms are related to segregations of certain natures (indicated by their codes below).

C	Concretions	Spheroidal mineral aggregates. Crudely concentric internal fabric can be seen with the naked eye. Includes pisoliths and ooliths (typically A, F, M, N and K in nature).
F	Fragments	Broken pieces of segregations.
G	Capsules	Solidified pupal cases of *Leptopius* spp. (typically *L. duponti*, Tilley *et al.* 1997). Also known as clogs. Typically A, K, F, N in nature.
H	Hard segregations	Irregular strong segregations that cannot be determined to be nodules or concretions (or fragments thereof). Typically L in nature but may have secondary cementation or encrustation.
L	Laminae	Planar, plate-like or sheet-like segregations. Typically M in nature.
M	Pseudomycelia	Fine to very fine traces with an appearance similar to fungal hyphae (usually white and K in nature, but should be tested to distinguish from biological material).
N	Nodules	Irregular, rounded mineral aggregates. No concentric or symmetric internal fabric. Can have hollow interior. Typically A, F, M, N and K in nature.
P	Powdery	Very fine, loose segregations, with a dust-like appearance. Typically K in nature.
R	Root linings	Linings of former or current root channels.
S	Soft segregations	Finely divided soft segregations. They contrast with surrounding soil in colour and composition but are not easily separated as discrete bodies. Boundaries may be clearly defined or diffuse. Typically K in nature.
T	Tubules	Medium or coarser (>2 mm wide) tube-like segregations, which may or may not be hollow. Typically K in nature.
U	Coralliferous	Strong segregations with a rough, irregular exterior and often a branched shape. Typically K in nature.

| V | *Veins* | Fine (<2 mm wide) linear segregations. Typically Y in nature. |
| X | *Crystals* | Single or complex clusters of crystals visible with the naked eye or a ×10 hand lens. Typically Y in nature. |

Size of segregations

Approximately equidimensional segregations (concretions, nodules) are measured in the greatest dimension. Segregations where one dimension is much greater than the other two (tubules, root linings, veins, laminae) are measured in the least dimension.

1	*Fine*	<2 mm
2	*Medium*	2–6 mm
3	*Coarse*	6–20 mm
4	*Very coarse*	20–60 mm
5	*Extremely coarse*	>60 mm

Strength of segregations

Strength may be recorded where appropriate.

| 1 | *Weak* | Can be broken between thumb and forefinger. |
| 2 | *Strong* | Cannot be broken between thumb and forefinger. |

Magnetic attributes of segregations

| M | *Magnetic* | Attracted onto surface of hand-held magnet. |
| N | *Non-magnetic* | Not attracted onto surface of hand-held magnet. |

Effervescence of segregations

Field tests for effervescence may be used to confirm the nature of some segregations – specifically 6% H_2O_2 for manganese and 1M HCl for calcium carbonate. The spot test for calcium carbonate reaction is described as method 19C1 in Rayment and Lyons (2011). See the Field Tests section (page 169) for further detail of these tests.

ROOTS

Record the presence of roots observed in each horizon in areas 100 mm^2 on a cleaned exposure face as per the categories listed below.

Root size

		Diameter
1	*Very fine*	<1 mm
2	*Fine*	1–2 mm
3	*Medium*	2–5 mm
4	*Coarse*	>5 mm

Root abundance

		Number of roots per 0.01 m² (100 mm × 100 mm)	
		Very fine and fine roots	Medium and coarse roots
0	*No roots*	0	0
1	*Few*	1–10	1–2
2	*Common*	10–25	2–5
3	*Many*	25–200	>5
4	*Abundant*	>200	>5

FIELD TESTS

A range of field tests may be used either singly or in conjunction. In all instances, the *method used* and the *depth* (in metres) at which the test was undertaken must be recorded. Some field methods are abbreviated or simplified versions of laboratory methods, consequently observers should be cognisant of the realistic precision and accuracy of field measurements and the difference between quantitative and qualitative field tests. Field tests may be undertaken on the soil matrix or as spot tests on specific components (e.g. segregations, see page 168).

Selection of points in the profile for field tests generally adopts either of two approaches:

- by horizons (usually the mid-point), or
- at selected intervals down the profile, where horizons are thick (greater than 0.3 m) and horizon boundaries are gradual/diffuse. This system gives useful data in some soils when important chemical changes may occur independently of visible horizon changes, for instance increasing salinity with depth in some cracking clays.

Field pH

pH is most commonly determined in the field using a field pH kit based on the specifications of Raupach and Tucker (1959). Using such a method, field pH may be *estimated to 0.5 of a unit*. Portable pH probes may also be used, but specific guidance should be sought from manufacturers regarding probe specifications for different operating environments (e.g. 1:5 soil:water mixture versus *in situ* mud). Instruments typically measure to 0.1 of a unit, but accuracy is strongly influenced by the method.

Electrical conductivity

Measurement of electrical conductivity in the field is possible using portable conductivity meters and the same 1:5 soil:water mixture used for the measurement of pH by probe. However, the method is more sensitive to variations in soil material (bulk density, clay content) than the measurement of pH and large differences to laboratory determined values can occur. Specific guidance should be sought from manufacturers regarding probe specifications for different operating environments.

Effervescence of carbonate in fine earth

The effervescence of the fine earth fraction (the soil matrix) is used as a qualitative indicator of 'finely divided carbonate', which may not be visible to the eye. The method (19D1) is described in detail in Rayment and Lyons (2011).

Determine the fine earth effervescence using sufficient 1M HCl (usually one to three drops).

N	*Non-calcareous*	No audible or visible effervescence (no bubbles).
S	*Slightly calcareous*	Slightly audible but no visible effervescence.
M	*Moderately calcareous*	Audible and slightly visible effervescence.
H	*Highly calcareous*	Moderate visible effervescence.
V	*Very highly calcareous*	Strong visible effervescence.

Hydrogen peroxide reaction

Hydrogen peroxide (H_2O_2) is used to test for the presence of manganese. An effervescent reaction is sought using 3–6% H_2O_2 in a manner similar to the use of HCl for testing carbonate (i.e. place a droplet on the segregation or soil in question). Effervescence indicates the presence of manganese oxide (MnO_2).

N	*Non-reactive*	No audible or visible effervescence of the fine earth fraction.
R	*Reactive*	Audible/visible effervescence of the fine earth fraction.

30% H_2O_2 (adjusted to pH 5.5) is used in acid sulfate soil field tests, to produce a stronger reaction and for the measurement of field oxidation (pH_{FOX}). Further information on acid sulfate soil methods should be obtained from the *National acid sulfate soils sampling and identification methods manual* (Sullivan *et al.* 2018) or any newer iteration of that document.

Water repellence

Water repellence of some soils, usually sandy, is caused by a series of long-chain polymethylene waxes, made up of acids, alcohols and esters, attached to the sand grains (Ma'shum *et al.* 1988). These soils occur Australia-wide but are widespread in southern Australia (McGhie and Posner 1980; Wetherby 1984) and the far north. Degree of repellence is assessed by determining the concentration of ethanol required to wet the sand in 10 seconds (King 1981). An abbreviated form of this method is recommended for routine field situations.

N	*Non water repellent*	Water is absorbed into soil in 10 seconds or less.
R	*Water repellent*	Water takes greater than 10 seconds and 2-molar ethanol takes 10 seconds or less to be absorbed into soil.
S	*Strongly water repellent*	2-molar ethanol takes greater than 10 seconds to be absorbed into soil.

Note: Soil temperature at testing should be between 15°C and 25°C. Higher temperatures will increase, and lower temperatures decrease, rates of absorption. Industrial-grade methylated spirits, available from chemists, at a concentration of 23.9 ml per 200 ml water can be substituted for the 2-molar ethanol to obtain approximate values.

Odour

Record soil odour (when appropriate) as per FAO (2006).

N	*None*	No odour detected.
P	*Petrochemical*	Presence of gaseous or liquid hydrocarbons (e.g. diesel, oil, petrol, bitumen).
S	*Sulfurous*	Presence of H_2S (hydrogen sulfide/rotten egg gas), commonly associated with strongly reduced soil containing sulfur compounds.

Slaking and dispersion

Qualitative slaking and dispersion tests of soil are normally undertaken in the laboratory (e.g. AS1289.3.8.1, Anon. 2017a; Loveday and Pyle 1973; Field *et al.* 1997; Emerson 2002). Some of these methods have been adapted for routine use in the field, usually with an abbreviated time period. Local analysis of the interactions of assessment time with dispersion behaviour of soils should be undertaken to provide confidence in data obtained using abbreviated methods.

SOIL WATER REGIME

Traditional approaches to soil drainage do not adequately differentiate between hydrological setting and permeability of the material of the profile. For example, a very permeable, coarse-textured soil occurring in a wet depression would have to be classed as very poorly drained. Hence there is a need to consider *permeability*, which refers to the potential of a soil to transmit water internally, and *drainage*, which refers to the rapidity and extent of water removal from the soil profile or site. Both these aspects of internal drainage may be difficult to assess in the field and cannot be based solely on profile morphology. Assessment is strongly influenced by the depth of exposure of the profile. Mottling may reflect drainage status, since mottling may be a relict feature. Vegetation and topography are likely to be useful guides.

The concepts of permeability and drainage given below are largely based on Canada Soil Survey Committee (1978). The terms have traditionally been applied to the profile as a whole, but it is possible to utilise the concepts (where appropriate) at the horizon scale. This is particularly relevant in soils where the upper profile may be highly permeable and/or freely drained and the lower profile is not, with consequential influence on the use of the soil for specific crops with differing rooting depths.

Soil permeability

Permeability is independent of climate and drainage, and – as applied to a soil – is controlled by the potential to transmit water (saturated hydraulic conductivity, Ks) of the least permeable layer in the soil. Therefore, it is influenced by, and inferred from, attributes of the soil such as mineralogy, structure, texture, porosity, cracks and shrink–swell properties. In the classes given below, the rate of transmission of water in the profile is based on the assumption that loss by evapotranspiration is minimal. The Ks ranges are compatible with those of Nowland in Canada, as reported by McKeague *et al.* (1982). Although Ks data are limited for Australian soils, values for some well-known soils may be found in Bonell *et al.* (1983), Talsma (1983), Williams (1983), Connolly *et al.* (1997), Ticehurst *et al.* (2007) and Smolinski *et al.* (2015).

1	*Very slowly permeable* Ks range: <5 mm/day (<1 mm/hr) Drainage time: months	Vertical transmission of water in the least permeable horizon is very slow; the profile would take periods of a month or more after thorough wetting to reach drained upper limit, if there were no obstructions to water movement within the profile. Structure may vary, but cracks or spaces between peds, when dry, close on wetting. Texture is usually clay or silty clay, and there are no interconnected pores visible (with a hand lens) that could conduct water when wet. Example soils include grey sodic clays and very strongly cemented pans.
2	*Slowly permeable* Ks range: 5–50 mm/day (<1–2 mm/hr) Drainage time: weeks	Vertical transmission of water in the least permeable horizon is slow; the profile would take periods of a week or more after thorough wetting to reach drained upper limit, if there were no obstructions to water movement within the profile. Structure may vary, from massive to strong grade. Texture is usually clay or silty clay, and there will be few interconnected pores visible (with a hand lens) that conduct water when wet. Vertic (2:1) clays are common. If texture is coarser, the interparticle voids are filled with fine mineral (e.g segregations or cementation). Example soils include Vertosols and strongly cemented pans.
3	*Moderately permeable* Ks range: 50–250 mm/day (2–21 mm/hr) Drainage time: days	Vertical transmission of water in the least permeable horizon is such that the profile would take no more than a number of days (1–5) after a thorough wetting to reach drained upper limit, if there were no obstructions to water movement from the profile. Texture is variable. The soil may be massive to structured. Structure grade is usually at least moderate, and prismatic or blocky peds are common. If massive, the soil material is always porous. The pores and channels that remain open when wet are clearly visible with a hand lens. Example soils include some Chromosols, some Dermosols, some Vertosols, some Kandosols and some pans.
4	*Highly permeable* Ks range: 250–500 mm/day (2–21 mm/hr) Drainage time: hours to days	Vertical transmission of water in the least permeable horizon is such that the profile would take no more than a number of hours (12–24) after a thorough wetting to reach drained upper limit, if there were no obstructions to water movement from the profile. The soil may vary in structure but grade is usually at least moderate, and blocky or polyhedral peds are common. If massive, the soil material is always porous. The pores and channels that remain open when wet are clearly visible with a hand lens. Example soils include red Kandosols, some Dermosols and some Chromosols.

| 5 | *Very highly permeable* Ks range: 500–1000 mm/day (21–42 mm/hr) Drainage time: hours | Vertical transmission of water in the least permeable horizon is such that the profile would take no more than a number of hours (1–12) after a thorough wetting to reach drained upper limit, if there were no obstructions to water movement from the profile. Horizons have large, continuous and clearly visible connecting pores and cracks that do not close with wetting. Texture may be sandy or clayey but clays are likely to only be 1:1 type. Sandy soils are likely to be apedal, but clay soils are likely to have strong polyhedral structure. Soil horizons are usually apedal, but some medium- to fine-textured soils with strong granular structure or cementation of aggregates can also be very highly permeable. Gravel or segregations may be an important contribution to permeability. Example soils include some sandy soils (Arenosols, Tenosols) and Ferrosols. |
| 6 | *Rapidly permeable* Ks range >1000 mm/day (>42 mm/hr) Drainage time: minutes to hours | Vertical transmission of water in the least permeable horizon is such that the profile would take no more than an hour after a thorough wetting to reach drained upper limit, if there were no obstructions to water movement from the profile. Horizons have large, continuous and clearly visible connecting pores. Macropore (bypass) flow may also be a key mechanism in the soil. Texture may be sandy or clayey but clays are likely to only be 1:1 type. Sandy soils are likely to be apedal, but clay soils are likely to have strong polyhedral structure. Example soils include sands (Arenosols) and red Ferrosols. |

Drainage

Drainage is a useful term to summarise local soil wetness conditions – it provides a statement about soil and site drainage likely to occur in most years. It is affected by a number of attributes, both internal and external, that may act separately or together. Internal attributes include soil structure, texture, porosity, hydraulic conductivity and water-holding capacity, while external attributes are source and quality of water, evapotranspiration, gradient and length of slope, and position in the landscape. Drainage is recorded as the site exists at the time of observation, within the site space. The nature of the drainage should therefore be recorded in conjunction with the drainage category.

Nature of drainage

N	*Natural drainage*	Drainage of the site has not been altered by anthropogenic activity.
A	*Artificially drained*	Drainage of the site has been increased by anthropogenic activity (e.g trenches, pits, mole drains, pipes).
W	*Artificially waterlogged*	Drainage of the site has been decreased through anthropogenic activity (e.g. banks, impoundment).

Drainage class

1	*Very poorly drained*	Water is removed from the soil so slowly that the watertable remains at or near the surface for most of the year. Surface flow, groundwater and subsurface flow are major sources of water, although precipitation may be important where there is a perched watertable and precipitation exceeds evapotranspiration. Soils have a wide range in texture and depth, and often occur in low-lying sites. Strong gleying and accumulation of surface organic matter are usually features of most soils.
2	*Poorly drained*	Water is removed very slowly in relation to supply. Subsurface and/or groundwater flow, as well as precipitation, may be a significant water source. Seasonal ponding, resulting from runon and insufficient outfall, also occurs. A perched watertable may be present. Soils have a wide range in texture and depth; many have horizons that are gleyed, mottled, or possess orange or rusty linings of root channels. All horizons remain wet for periods of several months.
3	*Imperfectly drained*	Water is removed only slowly in relation to supply. Precipitation is the main source if available water storage capacity is high, but subsurface flow and/or groundwater contribute as available water storage capacity decreases. Soils have a wide range in texture and depth. Some horizons may be mottled and/or have orange or rusty linings of root channels, and are wet for periods of several weeks.
4	*Moderately well-drained*	Water is removed from the soil somewhat slowly in relation to supply, due to low permeability, shallow watertable, lack of gradient, or some combination of these. Soils are usually medium to fine in texture. Significant additions of water by subsurface flow are necessary in coarse-textured soils. Some horizons may remain wet for as long as one week after water addition. Mottles may be present but typically in low abundance and not grey/pale.
5	*Well-drained*	Water is removed from the soil readily but not rapidly. Excess water flows downward readily into underlying, moderately permeable material or laterally as subsurface flow. The soils are often medium in texture. Some horizons may remain wet for several days after water addition.
6	*Rapidly drained*	Water is removed from the soil rapidly in relation to supply. Excess water flows downward rapidly if underlying material is highly permeable. There may be rapid subsurface lateral flow during heavy rainfall provided there is a steep gradient. Soils are usually coarse-textured or shallow, or both. No horizon is normally wet for more than several hours after water addition.

PARENT MATERIAL

The parent material of a soil profile may be recorded using one or more lithology codes from Table 35 in the Substrate chapter. Record confidence of the determination of parent material lithology using confidence codes in the Substrate chapter (page 178).

SAMPLING

Sampling of soil profiles is complex and influenced by many factors including microrelief, the nature of the soil profile, intended analysis and purpose of the investigation. Consideration of both horizontal and vertical variability is essential to correct sampling procedures. Sampling protocols are not described in detail here, but a number of key principles are worth noting:

- Record all necessary metadata concerning a sample (e.g. date, depth interval, sample number).
- Utilise all available field data to inform sample intervals and avoid sampling across horizon boundaries, particularly those that represent significant changes in soil properties (e.g. the A to B boundary in a texture contrast soil). A recognised exception to this is sampling of a standard interval of 0–0.1 m – in many parts of Australia, the A1 horizon is <0.1 m thick.
- Avoid samples that exceed 0.3 m in thickness.
- Understand the nature of horizon boundaries and the interaction with sample intervals when bulking across profiles.
- Analytical data from samples do not derive their full potential if other data are missing (e.g. profile morphology, land use).
- Ensure horizontal variability at a site is fully understood in the context of sampling (e.g. the influence of microrelief and, where relevant, mound/furrow or row/inter-row).
- Avoid bulking samples that come from an area that exceeds the dimension of the site.
- Avoid sampling areas that are not representative of the sampling framework (e.g. sampling disturbed areas when attempting to represent an undisturbed scenario).
- Ensure that sampling and handling methods are appropriate for the analysis required (e.g. the need to avoid certain types of lubricants or freeze samples).

PHOTOGRAPHY

The advent of digital photography has allowed for rapid acquisition of imagery of the soil profile, irrespective of how it has been exposed. Photographs of a soil profile should include a pit tape with metric divisions. It is also useful to photograph other key features, including the soil surface, vegetation and landscape. All necessary metadata should be recorded for each photograph (e.g. time/date, location, direction of view).

SUBSTRATE

JG Speight and RF Isbell

This chapter deals with materials and masses of earth or rock that underly the soil profile. They are not soils, but typically underlie them and may or may not be part of the *regolith*. The regolith is defined as 'everything from fresh rock to fresh air' (Eggleton 2001). The substrate includes the R layer and that part of the C layer that shows no pedological development (page 91), but excludes the solum (A and B horizons) and buried soil horizons, including D horizons and pans. The substrate beneath a soil profile may or may not be the parent material of the soil. While the determination of parent material can sometimes be difficult to ascertain, it can often be ascribed confidently.

The properties of the substrate should be described as objectively as possible and there is considerable overlap with regolith and geotechnical description (Anon. 2017b; Huggett 2023). The first group of properties refers to the *material* or substance in an intact state, as would be seen in a hand-sized specimen without cracks. Such properties serve to identify the type of rock, such as sandstone, or unconsolidated material, such as clay. A second group of properties comprising spacing of discontinuities, alteration and mass strength refers to substrate *masses*. These require observations of areas of greater dimensions. Types of substrate mass are classified mainly according to their inferred origin. Examples are alluvium, parna, ferricrete and saprolite. Texts such as Scott and Pain (2008) can be consulted for further detail regarding the description of regolith materials in Australia.

SUBSTRATE OBSERVATION

The substrate should be observed at the point of the soil profile observation or as close to it as may be practicable. Large vertical exposures of the substrate may reveal the spatial variation of substrate features.

Type of observation of substrate material

A	*Auger boring*
C	*Undisturbed soil core*
E	*Existing vertical exposure*
O	*Outcrop*, where presumed continuous with substrate
P	*Soil pit*

Distance

Estimate the distance in metres of the point of observation of substrate material from the point of soil observation.

Confidence that substrate is parent material

The observer should state the degree of confidence that the observed substrate material is in fact the *parent material* of the observed soil profile or the major part of that profile (i.e. of the B horizon).

A   *Almost certain or certain*

D   *Dubious, doubtful*

N   *Not parent material*

P   *Probable*

Depth

Measure or estimate, in metres, the depth of the point of observation of substrate material below the land surface.

PROPERTIES OF SUBSTRATE MATERIAL

The properties in this section refer to intact hand-sized samples or very small areas of outcrop. In many cases it will be necessary to use a ×10 hand lens to determine some of them.

Grain size

It is informative to estimate the size of the most common particles of a substrate material whether the material is thought to be of sedimentary, metamorphic or igneous origin.

1   *<0.06 mm*[29]   Silt- or clay-sized; grains not visible; for example, chert, shale, basalt, silt, coal.

2   *0.06–2 mm*   Sand-sized; grains visible; for example, sand, sandstone, dolerite, porphyry, sandy tuff, graywacke, micro-granite, schist, quartzite.

3   *>2 mm*   Gravel-sized; for example, gravel, conglomerate, breccia, pebbles, cobbles, stones, boulders, pegmatite, granite, agglomerate.

Texture

A   *Amorphous*   Without visible crystalline or fragmental texture, even through a hand lens.

F   *Fragmental*   Consisting of mineral or rock particles that are broken or abraded.

---

29   As shown by Figure 24, these size grade limits agree with the scale of Wentworth (1922), adopted by the United States National Research Council (Pettijohn 1957) and the American Geophysical Union (Lane et al. 1947), and with that of the Massachusetts Institute of Technology, also adopted by the British Standards Institution (1975) and the Standards Association of Australia (1977). The 0.06 mm boundary criterion (strictly 0.063 mm) separating sand and silt differs markedly from the value 0.02 mm of the International Scheme (proposed originally by Atterberg (1905) and accepted by the International Society of Soil Science); the 0.05 mm value of the United States Department of Agriculture (Soil Survey Staff 1975); and the 0.075 mm value of the Unified Soil Classification (Anon. 1953)

| P | *Porphyritic* | Crystalline, with individual larger crystals in a matrix of much smaller crystals or glass. |
| X | *Crystalline (non-porphyritic)* | Consisting of interlocking mineral crystals. |

Structure

B	*Bedded*	With planar surfaces marking successively deposited layers.
C	*Concretionary*	Spheroidal concretions cemented together.
F	*Fissile*	Easily split along closely spaced parallel planes.
L	*Foliated*	Consisting of sheets or laminae arranged in a layered manner.
P	*Platy*	Plate-like units cemented together.
R	*Vermicular*	Worm-like structure and/or cavities.
S	*Vesicular*	Sponge-like structure having large pores, which may or may not be filled with softer material.
V	*Massive*	No recognisable structure.

Porosity

0	*Non porous; dense*
1	*Slightly porous*
2	*Porous*

Mineral composition

Make provision for recording one dominant mineral and one or two minor minerals, as identified by inspection of the hand specimen.

C	*Carbonaceous material*
D	*Dark minerals*
F	*Feldspar*
G	*Glauconite*
K	*Carbonates (react with 1-molar HCl)*
L	*Clays (argillaceous)*
M	*Mica*
Q	*Quartz*
S	*Sesquioxides*
Y	*Gypsum*

Mottles and other colour patterns

Mottling and colour patterns in the substrate may be described using codes as per soil profiles on pages 136–138.

Strength of material

The strength of a specimen of soil substrate material may be crudely estimated in the field by striking it with the head-end or the pick-end of a geological hammer or by trying to cut it with a knife, and then referring to Table 32. These estimates refer to the unconfined (or uniaxial) compressive strength. The strength is that of the intact material rather than that of the mass, the strength of which has generally been reduced by the development of fractures and other phenomena.

Lithological type of substrate material

The properties above will key out many of the rock types and unconsolidated materials listed in Table 33 and Table 34.

Record the rock type only if it is definitely known or is confidently presumed. An alphabetical checklist for types of substrate material, both rocks and unconsolidated materials,

Table 32   Field estimation of strength class of intact rock material by cutting or striking with a knife, pick or hammer*

	Strength	Knife	Pick	Hammer (single blow)
VW	*Very weak rock* (1–25 MPa)	Deep cut	Crumbles	Flattened or powdered
W	*Weak rock* (25–50 MPa)	Shallow cut or scratch	Indents deeply	Shattered into many small fragments
M	*Moderately strong rock* (50–100 MPa)	Nil or slight mark	Indents shallowly	Breaks readily into a few large and some small fragments
S	*Strong rock* (100–200 MPa)	Nil	Nil	Breaks into one or two large fragments
VS	*Very strong rock* (>200 MPa)	Nil	Nil	Nil

* This table was developed through correspondence with MJ Selby (see Selby 1980; Piteau 1971, page 99 in Hoek and Bray 1977, and compare with Anon. 1977).

Table 33   Unconsolidated material* classification

		Non-volcanic				Volcanic	
Grain size class	**Diameter (mm)**	**Without significant carbonate**		**With significant carbonate**			
Very coarse grained	60	**BO** **SN** **CB**	Boulders Stones Cobbles			**BB**	Bombs (or blocks)
Coarse grained	2	**GV**	Gravel			**SK**	Scoria (or lapilli)
Medium grained	0.06[†]	**S**	Sand	**KS**	Calcareous sand	**AS**	Volcanic ash (sandy)
Fine grained	0.002	**Z**	Silt			**AF**	Volcanic ash (fine)
Very fine grained		**C**	Clay	**ML**	Marl		

* Material that is loose, plastic or does not exceed the strength of 'Very weak rock' in Table 32; can be dug with hand tools.
† See footnote 29.

Table 34   Rock type classification

Developed from a classification by Dearman (Anon. 1977) and that in BS 5930 1981 (British Standards Institution 1981)

Texture		Fragmented (cryptocrystalline or amorphous)					
Structure		Bedded					
Dominant mineral grains		Quartz, feldspar, rock fragments	Carbonate			Volcanic rock (juvenile)	Evaporite or organic matter
Grain size	Very coarse (Rudaceous) 2.0 mm	CONGLOMERATE (grains rounded) BRECCIA (grains angular)	LIMESTONE (CaCO$_3$) DOLOMITE (CaMg (CO$_3$)$_2$)	CALCIRUDITE		AGGLOMERATE (grains rounded) VOLCANIC BRECCIA (grains angular)	HALITE (NaCl) ANHYDRITE (CaSO$_4$) GYPSUM (CaSO$_4$, 2H$_2$O) COAL
	Coarse (Arenaceous) 0.06 mm			CALCARENITE		TUFF	
	Fine (Argillaceous) 0.002 mm	SILTSTONE	CALCAREOUS MUDSTONE	CALCILUTITE MARL			
	Very fine (Argillaceous)	CLAYSTONE					
	Amorphous or cryptocrystalline	CHERT, JASPER					
Genetic group		Sedimentary rocks					
		Detrital (Sd)				Pyroclastic (Sp)	Chemical (Sc)

Crystalline (or amorphous)							
Foliated		Massive					
Quartz, feldspar, mica	Various	Quartz, potassic and sodic feldspar	Potassic feldspar, (little quartz)	Sodic feldspar, dark minerals (little quartz)		Calcic feldspar, dark minerals	Dark minerals
GNEISS MIGMATITE	MARBLE (carbonate)	PEGMATITE				GABBRO	PYROXENITE (mainly pyroxene) PERIODOTITE (mainly olivine)
SCHIST	QUARTZITE GRANULITE HORNFELS AMPHIBOLITE SERPENTINE GREENSTONE	GRANITE ADAMELLITE GRANODIORITE	SYENITE	DIORITE			
PHYLLITE		MICROGRANITE APLITE QUARTZ PORPHYRY (porphyritic texture)	MICROSYENITE PORPHYRY (porphyritic texture)	MICRODIORITE		DOLERITE	
SLATE (strongly fissile)		RHYOLITE	TRACHYTE	ANDESITE		BASALT	
MYLONITE (intensely deformed)		VOLCANIC GLASS					
Metamorphic rocks (Me)		Igneous rocks (Ig)					
		Felsic		Mafic			Ultramafic

Table 35 Alphabetical list of lithological type of rocks and other materials*

	Lithology	Genetic type*		Lithology	Genetic type*
AB	Asbestos	(An)	**GS**	Gneiss	(Me)
AD	Adamellite	(Ig)	**GN**	Granite	(Ig)
AG	Agglomerate	(Sp)	**GD**	Granodiorite	(Ig)
AC	Alcrete (bauxite)	(Sc)	**GL**	Glass	(An)
AM	Amphibolite	(Me)	**GR**	Granulite	(Me)
AN	Andesite	(Ig)	**GV**	Gravel	(Uc)
AH	Anhydrite	(Sc)	**GW**	Graywacke	(Sd)
AP	Aplite	(Ig)	**GE**	Greenstone	(Me)
AR	Arkose	(Sd)	**GY**	Gypsum	(Sc)
AF	Ash (fine)	(Uc)	**HA**	Halite	(Sc)
AS	Ash (sandy)	(Uc)	**HO**	Hornfels	(Me)
AY	Ash (fly)	(An)	**IG**	Igneous rock	(Ig)
BA	Basalt	(Ig)		(unidentified)	
BB	Bombs (volcanic)	(Uc)	**JA**	Jasper	(Sc)
BI	Bitumen/asphalt	(An)	**LI**	Limestone	(Sd)
BR	Breccia	(Sd)	**MB**	Marble	(Me)
KA	Calcarenite	(Sd)	**ML**	Marl	(Uc)
KM	Calcareous mudstone	(Sd)	**ME**	Metamorphic rock	(Me)
KS	Calcareous sand	(Uc)		(unidentified)	
KL	Calcilutite	(Sd)	**MC**	Mica	(Ig)
KR	Calcirudite	(Sd)	**MD**	Microdiorite	(Ig)
KC	Calcrete	(Sc)	**MG**	Microgranite	(Ig)
CH	Chert	(Sc)	**MS**	Microsyenite	(Ig)
C	Clay	(Uc)	**MI**	Migmatite	(Me)
CF	Clay (fired)	(Uc)	**MU**	Mudstone	(Sd)
CO	Coal	(Sc)	**MY**	Mylonite	(Me)
CR	Coffee rock	(Sc)	**OB**	Obsidian	(Ig)
CG	Conglomerate	(Sd)	**PG**	Pegmatite	(Ig)
CT	Concrete	(An)	**PE**	Peridotite	(Ig)
CU	Consolidated rock		**PL**	Phonolite	(Ig)
	(unidentified)		**PH**	Phyllite	(Me)
CQ	Coquina	(Sd)	**PC**	Porcellanite	(Sc)
SD	Detrital sedimentary	(Sd)	**PO**	Porphyry	(Ig)
	rock (unidentified)		**PT**	Plastics and related materials	(An)
DI	Diorite	(Ig)	**PU**	Pumice	(Ig)
DR	Dolerite	(Ig)	**PY**	Pyroxenite	(Ig)
DM	Dolomite	(Sd)	**QZ**	Quartz	(Ig)
FC	Ferricrete	(Sc)	**QU**	Quartzite	(Me)
FS	Feldspar	(Ig)	**QP**	Quartz porphyry	(Ig)
GA	Gabbro	(Ig)	**QS**	Quartz sandstone	(Sd)

	Lithology	Genetic type*		Lithology	Genetic type*
RB	Red-brown hardpan	(Sc)	**ZS**	Siltstone	(Sd)
RH	Rhyolite	(Ig)	**SL**	Slate	(Sd)
S	Sand	(Uc)	**SY**	Syenite	(Me)
SA	Sandstone	(Sd)	**TO**	Tonalite	(Ig)
ST	Schist	(Me)	**TR**	Trachyte	(Ig)
SK	Scoria	(Uc)	**TU**	Tuff	(Ig)
SR	Serpentinite	(Ig)	**UC**	Unconsolidated material	(Uc)
SH	Shale	(Sd)		(unidentified)	
LC	Silcrete	(Sc)	**VB**	Volcanic breccia	(Sp)
Z	Silt	(Uc)	**VG**	Volcanic glass	(Ig)

* An – anthropogenic; Ig – igneous rocks; Me – metamorphic rocks; Sd – sedimentary rocks, detrital; Sc – sedimentary rocks, chemical or organic; Sp – sedimentary rocks, pyroclastic; Uc – unconsolidated material.

is given in Table 35. Only the more common materials are listed. Others can be recorded in free format. Note that some lithology types may exist as a substrate or a pan (see page 163), for example, calcrete and ferricrete. Expert judgement is required to determine whether such materials are pan or substrate.

GENETIC TYPE OF SUBSTRATE MASSES
Bedrock and regolith zones

The mantle of earth and rock, including *weathered rocks* and *sediments*, altered or formed by land surface processes is called the *regolith*. The underlying zone of rocks formed or altered by deep-seated crustal processes is the *bedrock*. Regolith and bedrock are regarded here as zones characterised by different processes, rather than as classes of material. The original definition of regolith by Merrill (1897) stresses the latter view and includes only unconsolidated materials.

The depth of the regolith zone ranges from zero, where bedrock outcrops at the surface, to over 100 m in areas of deep weathering (Ollier 1984). In areas without much sediment, the lower boundary of the regolith is the *weathering front* (Mabbutt 1961) where features due to weathering first appear. Where sediments are very thick, their lower layers become isolated from land surface processes by their depth and by reduced permeability due to compaction. Here the base of the regolith, which could be called a 'lithification front', is where most of the *sediments* transform to *sedimentary rocks*. Sedimentary rocks are often folded and faulted, but, at least in Australia's stable environment, most unconsolidated sediments remain flat-lying.

Masses within the regolith zone, in contrast with those within the bedrock zone, often have low density, very low strength and little cohesion between their particles or fragments. Despite Merrill's definition, not all materials follow this tendency. Strong and cohesive masses (e.g. ferricrete) may also be characteristic of the regolith. On the other hand, some layers of sedimentary rock never become strong. Other rocks are weakened within the bedrock zone by deep-seated processes. Table 36 assigns types of substrate mass to either the regolith zone or the bedrock zone.

Scheme of classification

Table 36 presents a scheme of classification of soil substrate masses as they are found in soil and land surveys. The main classes represent rock masses not yet weathered, those now being weathered, those transported and deposited but not yet consolidated, and those hardened while still near the surface.

Table 36   Genetic classification of substrate masses

Unweathered rocks of the bedrock zone	
IG	*Igneous rocks*
ME	*Metamorphic rocks*
PL	*Plutonic rocks*
SC	*Chemical and organic sedimentary rocks*
SD	*Detrital sedimentary rocks (including aeolianite)*
SP	*Pyroclastic rocks (including ignimbrite)*
SR	*Sedimentary rocks*
VO	*Volcanic rocks*
Weathered rocks	
DR	*Decomposed rock*
PW	*Partially weathered rock*
SA	*Saprolite*
Sediments (unconsolidated)	
AL	*Alluvium*
BE	*Beach sediment*
CD	*Creep deposit*
CO	*Colluvium*
ED	*Aeolian sediment*
ES	*Aeolian sand*
FI	*Fill*
GY	*Gypsum*
LA	*Lacustrine sediment*
LD	*Landslide deposit*
LO	*Loess*
MA	*Marine sediment*
MD	*Mudflow deposit*
PA	*Parna*
SE	*Scree*
SH	*Sheet flow deposit*
TI	*Till (glacial)*
VA	*Volcanic ash*
Masses hardened in the regolith	
• Chemically hardened materials	
AC	*Alcrete (bauxite)*
FC	*Ferricrete*
KC	*Calcrete*

Unweathered rocks of the bedrock zone	
LC	Silcrete
PC	Porcellanite
RB	Red-brown hardpan
• Evaporites	
EV	Other evaporites
GY	Gypsum
HA	Halite (rock salt)
• Artificially hardened and anthropogenic materials	
AT	Other artificially hardened materials
CN	Concrete
ST	Stabilised soil

Named classes in this table are defined in the following glossary. Since this is a genetic classification, diagnostic attributes may be hard to specify. Associated landforms should not be used as *recognition criteria* for observed substrate masses. They can be used to infer the nature of substrate masses that cannot be observed. In the same way, substrate observations should not be used as recognition criteria for landforms.

GLOSSARY OF SUBSTRATE MASS GENETIC TYPES

	Aeolianite	See *eolianite*.
AC	*Alcrete* (bauxite)	Indurated material rich in aluminium hydroxides. Commonly consists of cemented pisoliths and usually known as bauxite.
AL	*Alluvium*	Sediment mass deposited from transport by channelled stream flow or over-bank stream flow.
BE	*Beach sediment*	Sediment mass deposited from transport by waves or tides at the shore of a sea or lake.
CD	*Creep deposit*	*Colluvium* slowly displaced a short distance downslope as a result of small irregular movements, with the net movement increasing towards the land surface.
CN	*Concrete*	Artificial conglomerate rock mass of selected size grade material that has been hardened using Portland cement or other kinds of cement. Concrete is a weak rock (*ca.* 35 MPa unconfined compressive strength) usually reinforced with steel to increase its tensile strength.
CO	*Colluvium*	Sediment mass deposited from transport down a slope by gravity (*scree*), landslide (*landslide deposit*), mudflow (*mudflow deposit*), creep (*creep deposit*) or sheet flow (*sheet flow deposit*), but not by stream flow. Compared with alluvium, colluvium lacks bedding structure; is more variable in grain size (i.e. more poorly sorted); contains much local material; and is generally much more angular. Coarse particles may have particular alignments.

DR	*Decomposed rock*	Weathered material (typically soft and clay-rich) produced by thorough decomposition of rock masses due to exposure to land surface processes, but with no transport. It generally lacks any structures that may have been present in the unweathered rock (see also *saprolite* and *partially weathered rock*).
ED	*Eolian sediment*	Sediment mass deposited from transport by the wind.
ES	*Eolian sand*	*Eolian sediment* of sand size, often taking the form of dunes, with characteristic bedding structures.
ET	*Eolianite*	Consolidated *sedimentary rock* consisting of clastic material deposited by the wind.[30] Includes bioclastic calcarenites.
EV	*Evaporite*	Weak *sedimentary rock* or sediment formed by the precipitation of solutes from water bodies on the land surface, typically as *lacustrine sediments*. Includes *halite* and *gypsum*.
FC	*Ferricrete*	Indurated material rich in hydrated oxides of iron (usually goethite and hematite) occurring as cemented nodules and/or concretions, or as massive sheets. This material has been commonly referred to in local usage around Australia as laterite, duricrust or ironstone.
FI	*Fill*	Mass of artificial sediment formed by earth-moving works. Fill is sometimes compacted to the status of a very weak rock mass (*stabilised soil*), but typically remains an earth mass (Table 37). Garbage forms a very low-density, low-strength fill.
GY	*Gypsum*	*Evaporite* consisting of hydrated calcium sulphate. Non-hydrated calcium sulphate forms closely related masses called anhydrite. It may subsequently be transported by wind as fine crystals and form lunettes or more widespread sedimentary layers blanketing the landscape.
HA	*Halite* (rock salt)	*Evaporite* consisting of sodium chloride.
IG	*Igneous rocks*	Strong or very strong rock masses formed by solidification of molten rock matter (magma) derived from below the Earth's surface. The rocks are mainly composed of interlocking crystals. Types are distinguished in Table 34. *Plutonic rocks* and *volcanic rocks* are included.
IN	*Ignimbrite*	Very weak to strong *volcanic rock* mass deposited from a flow of ash, the stronger forms being welded together by residual heat during deposition.
KC	*Calcrete*	Any cemented, terrestrial carbonate accumulation that may vary significantly in morphology and degree of cementation. Also known as carbonate pan, calcareous pan, caliche, kunkar, secondary limestone, travertine. All show slight to strong effervescence with 1-molar HCl.

---

30   Bates and Jackson (1987)

LA	*Lacustrine sediment*	Sediment mass deposited from transport by waves and from sediment solution and suspension in still water in a closed depression on land.
LC	*Silcrete*	Strongly indurated siliceous material cemented by, and largely composed of, forms of silica, including quartz, chalcedony, opal and chert.
LD	*Landslide deposit*	*Colluvium* rapidly displaced many metres downslope by failure of a mass of earth or rock. If the mass is not already a part of the regolith, the landslide incorporates it in the regolith. Original rock structures are fragmented and disorganised by the action of the landslide.
LO	*Loess*	*Eolian sediment* of silt size.
MA	*Marine sediment*	Sediment mass deposited from transport by waves and from solution and suspension in sea water.
MD	*Mudflow deposit*	*Colluvium* mixed with water to form dense fluid, and rapidly displaced metres or kilometres downslope. The material is more thoroughly disaggregated than that of a *landslide deposit* and lacks the bedding and sorting of grain sizes seen in *alluvium*.
ME	*Metamorphic rocks*	Moderately strong to very strong rock masses formed by the chemical and physical alterations of igneous or sedimentary rocks under high temperatures and/or very high pressures within the Earth's crust. Types are distinguished in Table 34.
PA	*Parna*	Fine-grained calcareous *eolian sediment* consisting of 30–70% clay.
PC	*Porcellanite*	Dense argillaceous rock of varying degree of silicification with a conchoidal fracture and general appearance of unglazed porcelain.
PL	*Plutonic rocks*	*Igneous rocks* solidified at depth within the Earth's crust.
PW	*Partially weathered rock*	Weathered material produced by exposure of rock masses to land surface processes but with no transport. Partial decomposition results in changes in colour, texture, composition, strength or form of the parent rock mass (see also *decomposed rock* and *saprolite*).
RB	*Red-brown hardpan*	An informal name used for a particular indurated earthy material (see 'Pans', page 163). In many instances it is not known if the red-brown hardpan is a paleosol or a cemented sediment.[31]
SA	*Saprolite*	A particular form of *decomposed rock*. It is characterised by the preservation of structures (including 'texture' in the petrological sense) that were present in the unweathered rock.

---

31   See Wright (1983).

SC	*Chemical and organic sedimentary rocks*	*Sedimentary rocks* in which mineral grains or fragments are not important constituents. The group includes coal, chert and non-fragmental limestones as well as saline rocks (evaporites) such as halite (rock salt) and gypsum. Chemical and organic sedimentary rocks are common in the regolith zone.
SD	*Detrital sedimentary rocks*	*Sedimentary rocks* composed of mineral grains or fragments derived from pre-existing rocks. Types are distinguished in Table 34.
SE	*Scree*	*Colluvium* deposited after falling or rolling from cliffed or steep slopes, consisting of loose rock fragments of gravel size or larger, and generally lacking a fine interstitial component.
SH	*Sheet flow deposit*	*Colluvium* deposited from transport by a very shallow flow of water as a sheet, or network of rills on the land surface. Sheet flow deposits are very thin except at the foot of a slope and beneath sheet-flood fans.
SP	*Pyroclastic rocks*	*Sedimentary rocks* resulting from the deposition of airborne materials produced by volcanic eruptions.
SR	*Sedimentary rocks*	Weak or moderately strong rock masses formed by the hardening of sediments due to compaction, recrystallisation or cementation. These processes can occur within the regolith but are promoted by burial within the Earth's crust. Major categories of sedimentary rocks are detrital sedimentary rocks, pyroclastic rocks, and chemical and organic sedimentary rocks.
ST	*Stabilised soil*	Artificial mass with the strength grade of very weak rock. It results from the 'stabilisation' of an earth mass by a variety of processes: compaction; the admixture of lime, Portland cement, bitumen or other substances; heating; freezing; or electro-hardening. Cement stabilisation can produce a mass as strong as 10 MPa unconfined compressive strength.[32]
TI	*Till*	Sediment mass deposited from transport in ice, as in a glacier. Till is neither bedded nor sorted; it has a matrix of clay or silt enclosing larger particles of unweathered rock ranging up to large boulders.
VA	*Volcanic ash*	*Eolian sediment* consisting of relatively fine (<2 mm) pyroclastic material. It often contains a proportion of highly weatherable glass.
VO	*Volcanic rocks*	*Igneous rocks* solidified after eruption on to the land surface.

PROPERTIES OF SUBSTRATE MASSES

These properties generally require observation of a near-vertical face 1 m² or more in area.

---

32 Ingles and Metcalf (1972).

Spacing of discontinuities

Physical weathering opens up fissures or joints that reduce the strength of the rock mass relative to that of the intact rock material. These fissures generally increase in number towards the land surface. The following categories of discontinuity spacing apply (Deere 1968; Selby 1982):

S	*>3 m*	Solid; virtually unjointed.
M	*1–3 m*	Massive; few joints.
B	*300 mm–1 m*	Blocky; moderately jointed.
F	*50–300 mm*	Fractured; intensely jointed.
C	*<50 mm*	Crushed or shattered.

Alteration

Substrate materials (or coarse fragments) may be so extensively altered (as in deep weathering profiles) that it may be difficult or impossible to determine their original nature. Certain constituents may be either depleted or enriched. Thus, in many laterite profiles, some horizons are ferruginised, partially ferruginised and partially kaolinised, and the pallid zone kaolinised. Silicification may also be associated with deep weathering profiles although not exclusively so; for instance, some limestones may be variably silicified. In contrast, calcification, which is widespread in parts of southern Australia, is not usually associated with deep weathering. Wind erosion, leading to the creation of ventifacts, is considered a form of alteration.

B		Aluminium enriched.
F	*Ferruginised*	Iron enriched.
K	*Calcified*	Calcium carbonate enriched.
L	*Kaolinised*	Clay enriched, usually pale coloured (e.g. the pallid zone of a laterite profile).
S	*Silicified*	Silica enriched.
W	*Wind eroded*	Eroded by wind (in the case of ventifact coarse fragments).
O	*Other*	Deeply weathered but no specific nature.

In some instances, more than one type of alteration may be present, for example the mottled zone of a laterite profile may be both ferruginised and kaolinised. Where both types of alteration occur, record both.

Mass strength

The mass strength of bodies of earth or rock affects tree growth, land-forming processes and engineering works, but it is difficult to measure. Direct tests of mass strength are not proposed here. However, broad strength classes contribute to defining types of substrate mass.

Table 37 orders types of substrate mass in terms of their unconfined compressive strength, using the same strength classes as in Table 32. In engineering usage, masses with an unconfined compressive strength less than 1.0 MPa (or 1.25 MPa (Anon. 1977)) correspond to 'soil' or 'earth'. The engineering definitions of soil and rock are given by Terzaghi and Peck (1967): '*Soil* is a natural aggregate of mineral grains that can be separated by such gentle mechanical means as agitation in water. *Rock*, on the other hand, is a natural aggregate of minerals connected by strong and permanent cohesive forces.' In this context, 'soil' and 'earth' are synonyms (Standards Association of Australia 1981). Since 'soil' takes its pedological meaning throughout this Field Handbook, these low-strength substrate materials and masses are referred to as 'earth'. The grain size of earth material ranges from clay to gravel or larger fragments.

The geological distinction between sediments and sedimentary rocks occurs at about 25 MPa. This higher value is also appropriate for the minimum strength of the 'R (rock) layer' in

Table 37 Relative strength, density and seismic velocity of dry earth and rock masses in the regolith and bedrock zones*

Strength class	Unconfined compressive strength (MPa)	Bulk density (Mg/m³)	Seismic velocity† (m/s)	Zone Bedrock	Zone Regolith
	0.01	**1.3**	**240**		
Unconsolidated substrate masses: **E** Earth or 'soil'					Soil (Softer) saprolite Alluvium Colluvium Aeolian sediment Beach sediment Lacustrine sediment Marine sediment Fill
	1.0	1.8	600		
VW Very weak rock					Stabilised soil Till Evaporites (Harder) saprolite Highly weathered rock
	25	2.1	1500		
Consolidated (R horizon) substrate masses: **W** Weak rock				(Softer) sedimentary rocks (Softer) metamorphic rocks	Concrete Moderately weathered rock
	50	2.4	2000		
M Moderately strong rock				(Harder) sedimentary rocks (Softer) metamorphic rocks	Slightly weathered rock
	100	2.7	3000		
S or VS Strong or very strong rock				Igneous rocks (Harder) metamorphic rocks	Faintly weathered rock
	300	3.0	7000		

* See page 133. † Seismic velocity values given for weaker materials refer to the unsaturated state. Saturation with water may double the velocity.
Data sources: Bartlett (1971); CJ Braybrook (pers. comm.); Church (1981); Dobrin (1960); Hoek and Bray (1977); Kesel (1976); Polak and Pettifer (1976); Schmidt and Pierce (1976); Selby (1980, 1982); and FJ Taylor (pers. comm.).

soil profile description (page 129) that cannot be dug with hand tools. Table 37 distinguishes 'Unconsolidated substrate masses' from 'Consolidated (R layer) substrate masses' at the 25 MPa value. The table also shows corresponding values of bulk density and seismic velocity. At suitable sites, the seismic velocity of various subsurface layers can be measured using portable equipment (Williams 1988). Seismic velocity varies directly with mass strength because of a functional relation to elastic constants. Bulk density happens to vary in the same sense for most earth and rock masses. The value of any one attribute indicates the likely values of the other two.

For engineering works, the following broad generalisations can be made about the strength classes of Table 37. 'Earth' can be picked up and carried easily using earth-moving machines such as excavators and scrapers. When stronger materials are to be moved, the first step is to reduce their strength and density to that of 'earth'. Such material is too weak to form roads or dams without being artificially stabilised to the status of 'very weak rock' by compaction or other techniques (Ingles and Metcalf 1972). 'Very weak rock' can be dislodged with a bulldozer blade (or hand tools, for that matter), but it is easier to move if it is first broken up by a tractor-mounted ripper (Anon. 1983). 'Weak rock' must be ripped before it can be removed; this can be done using tractors weighing less than 40 tonnes (gross), such as the Caterpillar D8N (Anon. 1987a) and the Komatsu D155A (Anon. 1987b). 'Moderately strong rock' can be ripped by the heaviest tractors, but 'strong or very strong rock' can be broken only with explosives.

REFERENCES

ABARES (2016) The Australian Land Use and Management Classification Version 8. Australian Bureau of Agricultural and Resource Economics and Sciences, Canberra. https://www.agriculture.gov.au/abares/aclump/land-use/alum-classification.

Abed T, Stephens NC (2003) *Tree Measurement Manual for Farm Foresters: Practical Guidelines for Farm Foresters Undertaking Basic Tree Measurement in Farm Forest Plantations*. 2nd edn. (Ed. M Parsons) National Forest Inventory, Bureau of Rural Sciences, Canberra.

ACT Government (2015) *ACT Environmental Offsets Calculator Assessment Methodology*. Australian Capital Territory, Canberra. https://www.environment.act.gov.au/__data/assets/pdf_file/0005/728600/Schedule-1-Environmental-offsets-Assessment-Methodology-FINAL-2.pdf [Accessed 27 June 2022].

Anon. (1953) *Unified Soil Classification System*. US Army Corps of Engineers Waterways Experiment Station, Technical Manual 3–357. Vicksburg, Mississippi.

Anon. (1977) The description of rock masses for engineering purposes: report by the Geological Society Engineering Group Working Party. *Quarterly Journal of Engineering Geology* **10**, 355–388. doi:10.1144/GSL.QJEG.1977.010.04.01

Anon. (1983) *Handbook of Ripping*. 7th edn. Caterpillar Tractor Co., Peoria, Illinois.

Anon. (1987a) *Caterpillar Performance Handbook*. 18th edn. Caterpillar Inc., Peoria, Illinois.

Anon. (1987b) *Specifications and Application Handbook*. 10th edn. Komatsu Ltd, Tokyo.

Anon. (1994) World Resources 1994–1995. A Report by the World Resources Institute. Oxford University Press, New York.

Anon. (2017a) Australian Standard AS1289.3.8. Methods of testing soils for engineering purposes Soil classification tests – Dispersion – Determination of Emerson class number of a soil. Standards Australia.

Anon. (2017b) Australian Standard AS1726:2017. Geotechnical site investigations. Standards Australia.

Aquatic Ecosystems Task Group (2012a) Aquatic Ecosystems Toolkit. *Module 1: Aquatic Ecosystems Toolkit Guidance Paper*. Australian Government Department of Sustainability, Environment, Water, Population and Communities, Canberra.

Aquatic Ecosystems Task Group (2012b) Aquatic Ecosystems Toolkit. *Module 2. Interim Australian National Aquatic Ecosystem Classification Framework*. Australian Government Department of Sustainability, Environment, Water, Population and Communities, Canberra.

Atterberg A (1905) Die rationelle Klassifikation der Sande und Kiese. *Chemiker-Zeitung* **29**, 195–198.

Avery BW (1980) *Soil Classification for England and Wales*. Soil Survey Technical Monograph No. 14. Soil Survey of England and Wales, Harpenden, Herts.

Baines G, Webster M, Cook E, Johnston L, Seddon J (2013) *The vegetation of the Kowen, Majura and Jerrabomberra districts of the ACT*. Technical Report 28. Environment and Sustainable Development Directorate, ACT Government.

Bartlett AH (1971) *Geophysical aspects of engineering investigations*. ANZAAS Engineering Geology Symposium, Brisbane.

Bates RL, Jackson JA (Eds) (1987) *Glossary of Geology*. 3rd edn. American Geological Institute, Alexandria, Virginia.

Beadle NCW (1981) *The Vegetation of Australia*. Gustav Fisher, Stuttgart, Germany.

Beard JS, Beeston GR, Harvey JM, Hopkins AJM, Shepherd DP (2013) The vegetation of Western Australia at the 1:3 000 000 scale. *Conservation Science Western Australia* **9**, 1–152.

Beckmann GG, Hubble GD, Thompson CH (1970) Gilgai forms in north-east Australia. In *Symposium on Soils and Earth Structures in Arid Climates*. Adelaide pp. 88–93. The Institution of Engineers, Australia.

Bodí MB, Martin DA, Balfour VN, Santin C, Doerr SH, *et al.* (2014) Wildland fire ash: production, composition and eco-hydro-geomorphic effects. *Earth-Science Reviews* **130**, 103–127. doi:10.1016/j. earscirev.2013.12.007

Boenisch G, Kattge J (2019) TRY Plant Trait Database. https://www.try-db.org/TryWeb/Home.php [Accessed 05 April 2022].

Bonell M, Gilmour DA, Cassells DS (1983) A preliminary survey of the hydraulic properties of rainforest soils in tropical north-east Queensland and their implications for the runoff process. In *Rainfall Simulation, Runoff and Soil Erosion*. (Ed. J De Ploey) pp. 57–78. Catena Supplement No. 4. Schweizerbart Science Publishers, Stuttgart, Germany.

Brack C (1998) Forest mensuration database. https://fennerschool-associated.anu.edu.au/mensuration/BrackandWood1998/MENSHOME.HTM [Accessed 8/05/2024].

Brewer R (1960) Cutans: their definition, recognition and classification. *Journal of Soil Science* **11**, 280–292. doi:10.1111/j.1365-2389.1960.tb01085.x

Brewer R (1964) *Fabric and Mineral Analysis of Soils*. John Wiley and Sons, Inc.

British Standards Institution (1975) Methods of tests of soils for civil engineering purposes. BS 1377 Gr 10. British Standards Institution, London.

British Standards Institution (1981) Code of practice for site investigations. BS 5930: 1981. British Standards Institution, London.

Brocklehurst P, Lewis D, Napier D, Lynch D (2007) *Northern Territory Guidelines and Field Methodology for Vegetation Survey and Mapping*. Northern Territory Department of Natural Resources, Environment and the Arts, Darwin, NT, Australia. https://hdl.handle.net/10070/635994 [Accessed 28 June 2022].

Butler BE (1955) A system for the description of soil structure and consistence in the field. *Journal of the Australian Institute of Agricultural Science* **21**, 239–249.

Calders K, Adams J, Armston J, Harm B, Bauwens S, *et al.* (2020) Terrestrial laser scanning in forest ecology: expanding the horizon. *Remote Sensing of Environment* **251**(December), 112102. doi:10.1016/j.rse.2020.112102

Canada Soil Survey Committee (1978) *The Canada Soil Information System (Can SIS): Manual for Describing Soils in the Field*. Land Resource Research Institute, Ottawa, Ontario.

Capital Ecology (2018) *2017 Woodland Quality and Extent Mapping – ACT Government Environmental Offsets*. Prepared for Environment Offsets, ACT Parks and Conservation Service, Project no. 2756.

Carnahan JA (1977) Natural vegetation. In *Atlas of Australian Resources*. Second series.Department of Natural Resources, Canberra.

Carnahan JA (1986) Vegetation. In *The Natural Environment*. (Ed. DN Jeans) pp. 260–282. Sydney University Press, Sydney.

Casson N, Downes S, Harris A (2009) *Native Vegetation Condition Assessment and Monitoring Manual for Western Australia*. Department of Environment and Conservation, Perth.

Cayley A (1859) On contour and slope lines. *The London, Edinburgh and Dublin Philosophical Magazine and Journal of Science*, 4th Series, **18**, 264–268.

Chamberlin TC, Salisbury RD (1904) *Geology: Volume 1 – Geologic Processes and Their Results*. Henry Holt and Company, New York.

Cheal D (2010) *Growth stages and tolerable fire intervals for Victoria's native vegetation data sets. Fire and adaptive management*. Report no. 84.Victorian Government Depratment of Sustainability and Environment, Melbourne.

Church HK (1981) *Excavation Handbook*. McGraw-Hill, New York.

Commonwealth of Australia (2001) *A Directory of Important Wetlands in Australia*. 3rd edn. Environment Australia, Canberra.

Connolly RD, Freebairn DM, Bridge BJ (1997) Change in infiltration characteristics associated with cultivation history of soils in south-eastern Queensland. *Soil Research* **35**, 1341–1358. doi:10.1071/S97032

Corey B, Radford I, Carnes K, Hatherley E, Legge S (2013) *North-Kimberley Landscape Conservation Initiative: 2010-12 Performance Report*. Department of Parks and Wildlife, Kununurra, Western Australia.

Cowardin LM, Carter V, Golet FC, LaRoe ET (1979) Classification of wetlands and deepwater habitats of the United States. U.S. Fish and Wildlife Service. FWS/OBS-79/31. Washington, DC.

Crouch RJ (1976) Field tunnel erosion: a review. *Journal of the Soil Conservation Service of New South Wales* **32**, 98–111.

Cruden DM, Varnes DJ (1996) Landslide types and processes. In *Landslides: investigations and mitigation*. National Research Council, Transportation Research Board Special Report No. 247. (Eds AK Turner, RL Schuster). National Academy Press, Washington, DC.

CSIRO (2021) Identifying tropical rainforest plants. https://www.csiro.au/en/about/facilities-collections/Collections/ANH/TropicalRainforestID. [Accessed 10 January 2022].

Davis WM (1889) Topographic development of the Triassic formation of the Connecticut Valley. *American Journal of Science* **s3-37**, 423–434. doi:10.2475/ajs.s3-37.222.423

DAWE (2022) *Farm Forestry: Growing Together*. Department of Agriculture, Water and the Environment, Canberra.

De Cáceres M, Chytry M, Agrillo E, Attorre F, Botta-Duka't Z, *et al.* (2015) A comparative framework for broad-scale plot-based vegetation classification. *Applied Vegetation Science* **18**, 543–560. doi:10.1111/avsc.12179

Deere DV (1968) Geological considerations. In *Rock Mechanics in Engineering Practice*. (Eds OC Zienkiewicz, D Stagg) pp. 1–20. Wiley, New York.

Dengler J, Löbel S, Dolnik C (2009) Species constancy depends on plot size – a problem for vegetation classification and how it can be solved. *Journal of Vegetation Science* **20**, 754–766. doi:10.1111/j.1654-1103.2009.01073.x

Department of Environment and Resource Management (2011) Queensland Wetland Definition and Delineation Guideline. Queensland Government, Brisbane.

DES (2021) What are wetlands? Wetland*Info* website. Department of Environment and Science, Queensland. https://wetlandinfo.des.qld.gov.au/wetlands/what-are-wetlands/ [Accessed 19 September 2022].

Dobrin MB (1960) *Introduction to Geophysical Prospecting*. 2nd edn. McGraw-Hill, New York.

Donohue RJ, Harwood TD, Williams KJ, Ferrier S, McVicar TR (2014) *Estimating habitat condition using time series remote sensing and ecological survey data*. CSIRO Earth Observation and Informatics Transformational Capability Platform Client Report EP1311716. CSIRO, Canberra. doi:10.4225/08/58503778590e0

DPE (2022) *A revised classification of plant communities of eastern New South Wales*. Department of Planning and Environment, New South Wales Government.

DPIRD (2020) *Western Australian Rangeland Monitoring System (WARMS)*. https://www.agric.wa.gov.au/rangelands/western-australian-rangeland-monitoring-system-warms [Accessed 14 March 2022].

DSE (2004) *Vegetation Quality Assessment Manual – Guidelines for applying the habitat hectares scoring method*. Version 1.3. Victorian Government Department of Sustainability and Environment, Melbourne.

Dyne GR (1987) Two new Acanthodriline earthworms (Oligochaeta: Megascolecoidea) from the Northern Territory, Australia. *The Beagle, Records of the Northern Territory Museum of Arts and Sciences* **4**(1), 1–6.

Eggleton RA (2001) *The Regolith Glossary: Surficial Geology, Soils and Landscape*. CRC LEME, Canberra and Perth.

Emerson WW (2002) Emerson dispersion test. In *Soil Physical Measurement and Interpretation for Land Evaluation*. (Eds N McKenzie, K Coughlan, H Cresswell) pp. 190–199. CSIRO Publishing, Collingwood.

EPA (2016) *Technical Guidance: Flora and Vegetation Surveys for Environmental Impact Assessment*. Environment Protection Authority, Western Australia.

ESCAVI (Executive Steering Committee for Australian Vegetation Information) (2003) *Australian Vegetation Attribute Manual: National Vegetation Information System (Version 6.0)*. Department of Environment and Heritage, Canberra. https://www.dcceew.gov.au/sites/default/files/env/pages/06613354-b8a0-4a0e-801e-65b118a89a2f/files/vegetation-attribute-manual-6.pdf [Accessed 8/05/2024].

Eyre TJ, Kelly AL, Neldner VJ (2017) *Method for the Establishment and Survey of Reference Sites for BioCondition*. Version 3. Queensland Herbarium, Department of Science, Information Technology and Innovation, Brisbane.

Faber-Langendoen D, Keeler-Wolf T, Meidinger D, Tart D, Hoagland B, *et al.* (2014) EcoVeg: a new approach to vegetation description and classification. *Ecological Monographs* **84**, 533–561. doi:10.1890/13-2334.1

Falster D, Gallagher R, Wenk EH, Wright IJ, Indiarto D, *et al.* (2021) Austraits, a curated plant trait database for the Australian flora. *Scientific Data* **8**, 254. doi:10.1038/s41597-021-01006-6

FAO (2006) *Guidelines for Soil Description*. 4th edn. Soil Survey and Fertility Branch, Land and Water Division, Food and Agriculture Organization of the United Nations, Rome. https://www.fao.org/3/a0541e/a0541e.pdf

FAO (1988) 5. Classification. In *Nature and Management of Tropical Peat Soils*. FAO Soils Bulletin **59**. https://www.fao.org/3/x5872e/x5872e00.htm#Contents

Field DJ, McKenzie DC, Koppi AJ (1997) Development of an improved Vertisol stability test for SOILpak. *Soil Research* **35**(4), 843–852. doi:10.1071/S96118

Forest Practices Authority (2005) *Forest Botany Manual*. Modules 1 to 8. Forest Practices Authority, Hobart. https://www.fpa.tas.gov.au/Planning/biodiversity/forest_botany_manual [Accessed 5 May 2022].

Friedel MH, Chewings VH (1988) Comparison of crown cover estimates for woody vegetation in arid rangelands. *Australian Journal of Ecology* **13**(4), 463–468. doi:10.1111/j.1442-9993.1988.tb00994.x

Gallagher RV, Falster DS, Maitner BS, Salguero-Gómez R, Vandvik V, *et al.* (2020) Open Science principles for accelerating trait-based science across the Tree of Life. *Nature Ecology & Evolution* **4**, 294–303. doi:10.1038/s41559-020-1109-6

Gellie NJH, Hunter JT, Benson JS, Kirkpatrick JB, Cheal DC, *et al.* (2018) Overview of plot-based vegetation classification approaches within Australia. *Phytocoenologia* **48**, 251–272. doi:10.1127/phyto/2017/0173

Hallsworth EG, Robertson GK, Gibbons FR (1955) Studies in pedogenesis in New South Wales: VII. The 'gilgai' soils. *Journal of Soil Science* **6**(1), 1–31. doi:10.1111/j.1365-2389.1955.tb00826.x

Harms B (2023) To B and not B2 – the Australian soil horizon system: history and review. *Soil Research* **61**, 421–455. doi:10.1071/SR22154

Harwood TD, Donohue RJ, Williams KJ, Ferrier S, McVicar TR, *et al.* (2016) HCAS: a new way to assess the condition of natural habitats for terrestrial biodiversity across whole regions using remote sensing data. *Methods in Ecology and Evolution* **7**(9), 1050–1059. doi:10.1111/2041-210X.12579

Heard L, Channon B (1997) *Guide to a native vegetation survey using the biological survey of South Australia*. Information and Data Analysis Branch, Department of Housing and Urban Development, South Australian Government.

Hignett CT (2002) Measurement of soil strength using penetrometers. In *Soil Physical Measurement and Interpretation for Land Evaluation*. (Eds N McKenzie, K Coughlan, H Cresswell) Pp. 271–277. CSIRO Publishing, Melbourne.

Hnatiuk RJ, Thackway R, Walker J (2009) Vegetation. In *Australian Soil and Land Survey: Field Handbook*. 3rd edn. (Ed. National Committee on Soil and Terrain) pp. 73–125. CSIRO Publishing, Melbourne.

Hodgson JM (Ed.) (1974) *Soil Survey Field Handbook: Describing and Sampling Soil Profiles*. Soil Survey Technical Monograph No. 5. Soil Survey of England and Wales, Rothamsted Experimental Station, Harpenden, Herts.

Hoek E, Bray J (1977) *Rock Slope Engineering*. 2nd edn. The Institution of Mining and Metallurgy, London.

Hogg SE (1982) Sheetfloods, sheetwash, sheetflow or …? *Earth-Science Reviews* **18**(1), 59–76. doi:10.1016/0012-8252(82)90003-4

Horton RE (1945) Erosional development of streams and their drainage basins: hydrophysical approach to quantitative morphology. *Bulletin of the Geological Society of America* **56**, 275–370. doi:10.1130/0016-7606(1945)56[275:EDOSAT]2.0.CO;2

Houghton PD, Charman PEV (1986) *Glossary of Terms Used in Soil Conservation*. Soil Conservation Service of NSW, Sydney.

Huggett R (2023) Regolith or soil? An ongoing debate. *Geoderma* **432**, 116387. doi:10.1016/j.geoderma.2023.116387.

Ingles OG, Metcalf JB (1972) *Soil Stabilization: Principles and Practice*. Butterworths, Sydney.

Intergovernmental Committee on Surveying and Mapping (2002) *Geocentric datum of Australia technical manual*. Version 2.2. https://www.icsm.gov.au/sites/default/files/GDA2020TechnicalManualV1.1.1.pdf

International Society of Soil Science (1967) Proposal for a uniform system of soil horizon designations. *Bulletin – International Society of Soil Science* **31**, 4–7.

Isbell RF, NCST (2021) *The Australian Soil Classification*. 3rd edn. CSIRO Publishing, Melbourne.

IUSS Working Group (2022) *World Reference Base for Soil Resources: International soil classification system for naming soils and creating legends for soil maps*. 4th edn. International Union of Soil Sciences (IUSS), Vienna, Austria.

Jacobs MR (1955) *Growth habits of the eucalypts*. Forestry and Timber Bureau, Department of the Interior, Commonwealth Government Printer, Canberra.

Jarman SJ, Brown MJ, Kantvilas G (1984) *Rainforest in Tasmania*. Tasmanian National Parks and Wildlife Service, Hobart.

Jarman SJ, Kantvilas G, Brown MJ (1991) *Floristic and ecological studies in Tasmanian rainforest*. Tasmanian National Rainforest Conservation Program (NRCP) Australia, Report No. 3. Forestry Commission Tasmania and the Department of the Arts, Sport, the Environment, Tourism and Territories, Canberra.

Jarman SJ, Kantvilas G, Brown MJ (1999) Floristic composition of cool temperate rainforest. In *Vegetation of Tasmania. Flora of Australia Supplementary Series Number 8*. (Eds JB Reid, RS Hill, MJ Brown, MJ Hovenden) pp. 145–159. Australian Biological Resources Study, Hobart.

Jennings JN, Mabbutt JA (1977) Physiographic outlines and regions. In *Australia: A Geography*. (Ed. DN Jeans) pp. 38–52. Sydney University Press, Sydney.

Keith DA (2017) *Australian Vegetation*. Cambridge University Press, Melbourne.

Keith DA, Gorrod E (2006) The meanings of vegetation condition. *Ecological Management & Restoration* **7**, S7–S9. doi:10.1111/j.1442-8903.2006.00285.x

Keith DA, Ferrer-Paris JR, Nicholson E, Kingsford RT (Eds) (2020) The IUCN Global Ecosystem Typology 2.0: Descriptive profiles for biomes and ecosystem functional groups. Gland, Switzerland, IUCN. doi:10.2305/IUCN.CH.2020.13.en

Kenneally KF (2018) Kimberley tropical monsoon rainforests of Western Australia: perspectives on biological diversity. *Journal of the Botanical Research Institute of Texas* **12**, 149–228. doi:10.17348/jbrit.v12.i1.927

Kesel RH (1976) The use of refraction-seismic techniques in geomorphology. *Catena* **3**, 91–98. doi:10.1016/S0341-8162(76)80020-3

King LC (1953) Canons of landscape evolution. *Bulletin of the Geological Society of America* **64**, 721–752. doi:10.1130/0016-7606(1953)64[721:COLE]2.0.CO;2

King PM (1981) Comparison of methods for measuring severity of water-repellency of sandy soils and assessment of some factors that affect its measurement. *Australian Journal of Soil Research* **19**, 275–285. doi:10.1071/SR9810275

Kitchener A, Harris S (2013) *From Forest to Fjaeldmark: Descriptions of Tasmania's Vegetation*. Edition 2. Department of Primary Industries, Parks, Water and Environment, Hobart. https://nre.tas.gov.au/conservation/flora-of-tasmania/from-forest-to-fjaeldmark-descriptions-of-tasmanias-vegetation [Accessed 5 May 2022].

Lane EW, Brown C, Gibson GC, Howard CS, Krumbein WC, *et al.* (1947) Report of the subcommittee on sediment terminology. *Transactions - American Geophysical Union* **28**, 936–938. doi:10.1029/TR028i006p00936

Lange RT, Sparrow AD (1992) Growth rates of western myall (*Acacia papyrocarpa* Benth.) during its main phase of canopy spreading. *Australian Journal of Ecology* **17**, 315–320. doi:10.1111/j.1442-9993.1992.tb00813.x

Lawrie JW (1978) Hardpans in western New South Wales. In *Proceedings 1st Int. Rangeland Conference*. pp. 303–306. International Rangelands Society, Denver, Colorado.

Laws M, McCallum K, Bignall J, Kilpatrick E, O'Neill S, Sparrow B (2023a) Basal Area Module. In *Ecological Field Monitoring Protocols Manual Using the Ecological Monitoring System Australia*. (Eds S O'Neill, K Irvine, A Tokmakoff, B Sparrow). TERN, Adelaide.

Laws M, Morgan R, McCallum K, Potter T, Cox B, *et al.* (2023b) Cover Module. In *Ecological Field Monitoring Protocols Manual Using the Ecological Monitoring System Australia*. (Eds S O'Neill, K Irvine, A Tokmakoff, B Sparrow). TERN, Adelaide.

Lewis D, Patykowski J, Nano C (2021) Comparing data subsets and transformations for reproducing an expert-based vegetation classification of an Australian tropical savanna. *Australian Journal of Botany* **69**(7), 423–435. doi:10.1071/BT20164

Liang S, Wang J (2020) Chapter 12 – Fractional vegetation cover. In *Advanced Remote Sensing*. 2nd edn. pp. 477–510. Academic Press. https://www.sciencedirect.com/science/article/pii/B978012815826500012X

Löffler E (1974) Geomorphology of Papua New Guinea (map). CSIRO Australia Land Research Series No. 33. Canberra.

Löffler E, Ruxton BP (1969) Relief and landform map of Australia. In *The Representative Basin Concept in Australia*. Australian Water Resources Council, Hydrological Series No. 2. Canberra.

Loveday J, Pyle JC (1973) The Emerson dispersion test and its relationship to hydraulic conductivity. CSIRO Australia Division of Soils Technical Paper No. 15. Melbourne.

Lucas R, Van De Kerchove R, Otero V, Lagomasino D, Fatoyinbo L, *et al.* (2020) Structural characterisation of mangrove forests achieved through combining multiple sources of remote sensing. *Remote Sensing of Environment* **237**, 111543. doi:10.1016/j.rse.2019.111543

Luxton S, Lewis D, Chalwell S, Addicott E, Hunter J (2021) Australian advances in vegetation classification and the need for a national, science-based approach. *Australian Journal of Botany* **69**, 329–338. doi:10.1071/BT21102

Ma'shum M, Tate ME, Jones GP, Oades JM (1988) Extraction and characterisation of water-repellent materials from Australian soils. *Journal of Soil Science* **39**, 99–110. doi:10.1111/j.1365-2389.1988.tb01198.x

Mabbutt JA (1961) 'Basal surface' or 'Weathering front'. *Proceedings of the Geologists' Association* **72**, 357–358. doi:10.1016/S0016-7878(61)80019-9

Macvicar CN (1969) A basis for the classification of soil. *Journal of Soil Science* **20**, 141–152. doi:10.1111/j.1365-2389.1969.tb01563.x

Mark DM (1974) Line intersection method for estimating drainage density. *Geology* **2**, 235–236. doi:10.1130/0091-7613(1974)2<235:LIMFED>2.0.CO;2

Marshall TJ (1947) *Mechanical composition of soil in relation to field descriptions of texture*. CSIR Australia Bulletin No. 224. Melbourne.

McCoy RM (1971) Rapid measurement of drainage density. *Bulletin of the Geological Society of America* **82**, 757–762. doi:10.1130/0016-7606(1971)82[757:RMODD]2.0.CO;2

McDonald RC (1977) *Soil horizon nomenclature*. Queensland Department of Primary Industries Agricultural Chemistry Branch Technical Memorandum 1/77.

McElroy CT (1952) Contour trench formations in upland plains of New South Wales. *Journal and Proceedings of the Royal Society of New South Wales* **85**, 53–63.

McGhie DA, Posner AM (1980) Water-repellence of a heavy textured Western Australian surface soil. *Australian Journal of Soil Research* **18**, 309–323. doi:10.1071/SR9800309

McKeague JA, Wang C, Topp GC (1982) Estimating saturated hydraulic conductivity from soil morphology. *Soil Science Society of America Journal* **46**, 1239–1244. doi:10.2136/sssaj1982.03615995004600060024x

McKenzie NJ, Grundy MJ (2008) 2. Approaches to land resource survey. In *Guidelines for Surveying Soil and Land Resources*. 2nd edn. (Eds NJ McKenzie, MJ Grundy, R Webster, AJ Ringrose-Voase) pp. 15–26. CSIRO Publishing, Collingwood.

McKenzie NJ, McDonald WS, Murtha GG (1995) *Network of Australian soil and land reference sites. Design and specification*. ACLEP Technical Report No. 1. Australian Collaborative Land Evaluation Program. CSIRO Division of Soils, Canberra.

McKenzie NJ, Ringrose-Voase AJ, Grundy MJ (2008) *Guidelines for Surveying Soil and Land Resources*. 2nd edn. CSIRO Publishing, Collingwood.

McKenzie NL (1991) An ecological survey of tropical rainforests in Western Australia: Background and methods. In *Kimberley Rainforests Australia*. (Eds NL McKenzie, RB Johnston, PG Kendrick) pp. 1–26. Surrey Beatty and Sons, Sydney.

Meagher D (1991) *The MacMillan Dictionary of The Australian Environment*. The MacMillan Company of Australia, Melbourne.

Melton FA (1936) An empirical classification of flood-plain streams. *Geographical Review* **26**, 593–609. doi:10.2307/209717

Merrill GP (1897) *A Treatise on Rocks: Rock-weathering and Soils*. Macmillan, New York.

Michaels K, Panek D, Kitchener A, Contributing Eds (2020) *TASVEG VCA Manual: A manual for assessing vegetation condition in Tasmania*. Version 2.0. Natural and Cultural Heritage, Department of Primary Industries, Parks, Water and Environment, Hobart.

Moore ID, O'Loughlin EM, Burch GJ (1988) A contour based topographic model for hydrological and ecological applications. *Earth Surface Processes and Landforms* **13**, 305–320. doi:10.1002/esp.3290130404

Morse RJ, Craze B, Atkinson G, Crichton JR, Ryan PT, *et al.* (1987) *New South Wales soil data system handbook*. Draft edn. Soil Conservation Service of NSW, Sydney.

Mucina L (1997) Classification of vegetation: Past, present and future. *Journal of Vegetation Science* **8**, 751–760. doi:10.2307/3237019

Muir J, Shmidt M, Tindall D, Trevithick R, Scarth P, Stewart JB (2011) *Field measurement of fractional ground cover: a technical handbook supporting ground cover monitoring for Australia*. Department of Environment and Resource Management for the Australian Bureau of Agricultural and Resource Economics and Sciences, Canberra.

Native Vegetation Council (2020a) *Bushland Assessment Manual*. Native Vegetation Management Unit, Government of South Australia.

Native Vegetation Council (2020b) *Rangelands Assessment Manual*. Native Vegetation Management Unit, Government of South Australia.

Native Vegetation Council (2020c) *Scattered Tree Assessment Manual*. Native Vegetation Management Unit, Government of South Australia.

Natural and Cultural Heritage Division (2015) *Guidelines for Natural Values Surveys – Terrestrial Development Proposals*. Department of Primary Industries, Parks, Water and Environment. Tasmanian Government, Hobart.

Neldner V, Butler D (2008) Is 500m² an effective plot size to sample floristic diversity for Queensland's vegetation? *Cunninghamia* **10**, 513–519.

Neldner VJ, Wilson BA, Dillewaard HA, Ryan TS, Butler DW, *et al.* (2022) *Methodology for survey and mapping of regional ecosystems and vegetation communities in Queensland*. Version 6.0. Queensland Herbarium. Queensland Department of Environment and Science, Brisbane.

NLWRA (National Land and Water Resources Audit) (2001) Rangelands - Tracking Changes - Australian Collaborative Rangeland Information System. https://catalogue.nla.gov.au/catalog/1707659.

Northcote KH (1979) *A Factual Key for the Recognition of Australian Soils*. 4th edn. Rellim Technical Publications, Glenside, SA.

NVIS Technical Working Group (2017) *Australian Vegetation Attribute Manual: National Vegetation Information System, Version 7.0*. Department of the Environment and Energy, Canberra.

Ollier C (1984) *Weathering*. 2nd edn. Oliver and Boyd, Edinburgh.

Oyama M, Takehara H (1970) *Revised standard soil color charts*. Frank McCarthy Color Pty Ltd, Melbourne.

Pain C, Gregory L, Wilson P, McKenzie N (2011) The physiographic regions of Australia – Explanatory notes 2011. Australian Collaborative Land Evaluation Program and National Committee on Soil and Terrain. https://www.clw.csiro.au/aclep/documents/PhysiographicRegions_2011.pdf

Parkes D, Newell G, Cheal D (2003) Assessing the quality of native vegetation: the 'habitat hectares' approach. *Ecological Management & Restoration* **4**, s29–s38. doi:10.1046/j.1442-8903.4.s.4.x

Paton TR (1974) Origin and terminology for gilgai in Australia. *Geoderma* **11**(3), 221–242. doi:10.1016/0016-7061(74)90019-6

Patykowski J, Cowie I, Cuff N, Chong C, Nano C, *et al.* (2021) Can sampling for vegetation characterisation surrogate for species richness? Case studies from the wet–dry tropics of northern Australia. *Australian Journal of Botany* **69**(7), 375–385. doi:10.1071/BT20158

Penridge LK, Walker J (1988) The crown-gap ratio (C) and crown cover: derivation and simulation study. *Australian Journal of Ecology* **13**, 109–120. doi:10.1111/j.1442-9993.1988.tb01420.x

Pettijohn FJ (1957) *Sedimentary Rocks*. 2nd edn. Harper, New York.

Piteau DR (1971) Geological factors significant to the stability of slopes cut in rock. In *Planning Open Pit Mines*. (Ed. PWJ van Rensburg) pp. 33–53. AA Balkema, Amsterdam.

Polak EJ, Pettifer GR (1976) The use of surface geophysical methods in underground water investigations: proceedings of a symposium by the Australian Water Resources Council, Adelaide, August 1975. (Eds EJ Polak and GR Pettifer) Department of National Resources, Bureau of Mineral Resources, Geology and Geophysics Record No. 108. https://pid.geoscience.gov.au/dataset/ga/13528

Powell B (2008) 19. Classifying soil and land. In *Guidelines for Surveying Soil and Land Resources*. 2nd edn. (Eds NJ McKenzie, MJ Grundy, R Webster, AJ Ringroase-Voase) pp. 307–315. CSIRO Publishing, Collingwood.

Price O, Brocklehurst P (2008) *Quality assessment manual for native vegetation in the Northern Territory of Australia: A: Vegetation condition assessment 1. Top End forests and woodlands*. Department of Natural Resources, Environment, the Arts and Sport, Northern Territory Government, Palmerston. https://hdl.handle.net/10070/673751 [Accessed 28 June 2022].

Raupach M, Tucker BM (1959) The field determination of soil reaction. *Journal of the Australian Institute of Agricultural Science* **25**, 129–133.

Rayment GE, Lyons DJ (2011) *Soil Chemical Methods: Australasia*. CSIRO Publishing, Collingwood.

Reid JB, Hill RS, Brown MJ, Hovenden MJ (Eds) (1999) *Vegetation of Tasmania*. Flora of Australia Supplementary Series Number 8. University of Tasmania, Forestry Tasmania, CRC for Sustainable Production Forestry, Hobart.

Russell-Smith J (1991) Classification, species richness, and environmental relations of monsoon rain forest in Northern Australia. *Journal of Vegetation Science* **2**, 259–278. doi:10.2307/3235959

Schmidt PW, Pierce KL (1976) Mapping of mountain soils west of Denver, Colorado, for landuse planning. In *Geomorphology and Engineering*. (Ed. DR Coates) pp. 43–54. Dowden, Hutchinson and Ross, Stroudsburg, Pennsylvania.

Schoeneberger PJ, Wysocki DA (2017) *Geomorphic description system, version 5*. Natural Resources Conservation Service, National Soil Survey Center, Lincoln, NE. https://www.nrcs.usda.gov/resources/guides-and-instructions/geomorphic-description-system

Schoeneberger, PJ, Wysocki DA, Benham EC, Soil Survey Staff (2012) *Field book for describing and sampling soils, Version 3.0*. Natural Resources Conservation Service, National Soil Survey Center, Lincoln, NE.

Scott K, Pain C (2008) *Regolith Science*. CSIRO Publishing, Collingwood.

Selby MJ (1980) A rock mass strength classification for geomorphic purposes: with tests from Antarctica and New Zealand. *Zeitschrift für Geomorphologie* **24**, 31–51. doi:10.1127/zfg/24/1984/31

Selby MJ (1982) *Hillslope Materials and Processes*. Oxford University Press, Oxford.

Shien PT, Seneviratne HN, Ismail DSA (2011) A study on factors influencing the determination of moisture content of fibrous peat. *Journal of Civil Engineering, Science and Technology* 2(2), 39–47.

Sivertsen D (2009) *Native Vegetation Interim Type Standard*. Department of Environment, Climate Change and Water NSW, Sydney.

Smith GD, Arya LM, Stark J (1975) The densipan, a diagnostic horizon of densiaquults for soil taxonomy. *Soil Science Society of America Proceedings* 39, 369–370. doi:10.2136/sssaj1975.03615995003900020036x

Smolinski H, Pathan S, Galloway P, Kuswardiyanto K, Laycock J (2015) *Cockatoo Sands in the Victoria Highway and Carlton Hill areas, East Kimberley: land capability assessment for developing irrigated agriculture*. Resource management technical report 391. Department of Agriculture and Food, Western Australia, Perth.

So HB, Smith GD, Raine SR, Schafer BM, Loch RJ (Eds) (1994) Sealing, crusting and hardsetting soils: productivity and conservation. *Proceedings of the Second International Symposium on sealing, crusting and hardsetting soils: productivity and conservation*. University of Queensland, Brisbane, Australia 7–11 February 1994. Australian Society of Soil Science Inc. (Queensland Branch).

Soil Science Division Staff (2017) *Soil Survey Manual*. USDA Agricultural Handbook No. 18. Government Printer, Washington, DC. https://www.nrcs.usda.gov/sites/default/files/2022-09/The-Soil-Survey-Manual.pdf

Soil Survey Staff (1951) *Soil Survey Manual*. USDA Agricultural Handbook No. 18. Government Printer, Washington, DC.

Soil Survey Staff (1975) *Soil taxonomy: a basic system of soil classification for making and interpreting soil surveys*. USDA Agricultural Handbook No. 436. Government Printer, Washington, DC.

Sparrow BD, Foulkes JN, Wardle GM, Leitch EJ, Caddy-Retalic S, *et al.* (2020) A vegetation and soil survey method for surveillance monitoring of rangeland environments. *Frontiers in Ecology and Evolution* 8, 157–174. doi:10.3389/fevo.2020.00157

Specht RL (1970) Vegetation. In *Australian Environment*. (Ed. GW Leeper) pp. 44–67. Melbourne University Press, Melbourne.

Specht RL, Roe EM, Boughton VH (Eds) (1974) Conservation of major plant communities in Australia and Papua New Guinea. *Australian Journal of Botany Supplement* No. 7, 1–667.

Specht RL, Specht A, Whelan MB, Hegarty EE (1995) *Conservation Atlas of Plant Communities in Australia*. Southern Cross University Press, Lismore, NSW.

Speight JG (1967) Appendix 1: explanation of land system descriptions. In *Lands of Bougainville and Buka Islands, Territory of Papua and New Guinea*. (Eds RM Scott, PB Heyligers, JR McAlpine, JC Saunders, JG Speight) pp. 174–184. CSIRO Australia Land Research Series No. 20.

Speight JG (1971) Log–normality of slope distributions. *Zeitschrift für Geomorphologie* 15, 290–311.

Speight JG (1974) A parametric approach to landform regions. Institute of British Geographers Special Publication No. 7, 213–230.

Speight JG (1976) Numerical classification of landform elements from air photo data. *Zeitschrift für Geomorphologie* Suppl. 25, 154–168.

Speight JG (1977) Landform pattern descriptions from aerial photographs. *Photogrammetria* 32, 161–182. doi:10.1016/0031-8663(77)90012-6

Speight JG (1980) The role of topography in controlling through-flow generation: a discussion. *Earth Surface Processes and Landforms* 5, 187–191. doi:10.1002/esp.3760050209

Standards Association of Australia (1977) *Australian standard 1289: methods of testing soils for engineering purposes*. Standards Association of Australia, Sydney.

Standards Association of Australia (1981) *Site investigations, known as the SAA Site Investigation Code, AS 1726–1981*. Standards Association of Australia, Sydney.

Sullivan L, Ward N, Toppler N, Lancaster G (2018) *National Acid Sulfate Soils guidance: National acid sulfate soils sampling and identification methods manual.* Department of Agriculture and Water Resources, Canberra.

Sun D, Hnatiuk R, Neldner V (1997) Review of vegetation classification and mapping systems undertaken by major forested land management agencies in Australia. *Australian Journal of Botany* **45**, 929–948. doi:10.1071/BT96121

Talsma T (1983) Soils of the Cotter catchment area, ACT: distribution, chemical and physical properties. *Australian Journal of Soil Research* **21**, 241–255. doi:10.1071/SR9830241

TERN (2023) Ecological Monitoring System Australia. https://www.tern.org.au/emsa-protocols-manual [Accessed 2/05/2024].

Terzaghi K, Peck RB (1967) *Soil Mechanics in Engineering Practice.* 2nd edn. Wiley, New York.

Thackway R, Neldner VJ, Bolton MP (2008) Vegetation. In *Australian Soil and Land Survey Handbook: Guidelines for Surveying Soil and Land Resources.* (Eds NJ McKenzie, MJ Grundy, R Webster, AJ Ringrose-Voase) pp. 115–142. CSIRO Publishing, Melbourne.

Thwaites RN, Brooks AP, Pietsch TJ, Spencer JR (2022) What type of gully is that? The need for a classification of gullies. *Earth Surface Processes and Landforms* **47**(1), 109–128. doi:10.1002/esp.5291

Thwaites RT, Brooks A, Spencer J (in press) *The Queensland Gully Classification 2.2 Handbook (version 1).* Griffith University and The Queensland Government.

Ticehurst JL, Cresswell HP, McKenzie NJ, Glover MR (2007) Interpreting soil and topographic properties to conceptualise hillslope hydrology. *Geoderma* **137**, 279–292. doi:10.1016/j.geoderma.2006.06.016

Tilley DB, Barrows TT, Zimmerman EC (1997) Bauxitic insect pupal cases from northern Australia. *Alcheringa: An Australasian Journal of Palaeontology* **21**(2), 157–160. doi:10.1080/03115519708619182

Tomlinson M, Boulton AJ (2010) Ecology and management of subsurface groundwater dependent ecosystems in Australia – a review. *Marine and Freshwater Research* **61**, 936–949.

Twidale CR (1981) Origins and environments of pediments. *Journal of the Geological Society of Australia* **28**, 423–434. doi:10.1080/00167618108729179

Twidale CR (2014) Pediments and platforms: problems and solutions. *Géomorphologie* **20**(1), 43–56. doi:10.4000/geomorphologie.10480

Walker J, Hopkins MS (1990) Vegetation. In *Australian Soil and Land Survey Handbook: Field Handbook.* 2nd edn. (Eds RC McDonald, RF Isbell, JG Speight, J Walker, MS Hopkins) pp. 58–86. Inkata Press, Melbourne.

Walker J, Crapper PF, Penridge LK (1988) The crown-gap ratio (C) and crown cover: the field study. *Australian Journal of Ecology* **13**, 101–108. doi:10.1111/j.1442-9993.1988.tb01419.x

Warren JF (1965) The scalds of western New South Wales – a form of water erosion. *The Australian Geographer* **9**, 282–292. doi:10.1080/00049186508702435

Warren MW, Kauffman JB, Murdiyarso D, Anshari G, Hergoualc'h K, *et al.* (2012) A cost-efficient method to assess carbon stocks in tropical peat soil. *Biogeosciences* **9**, 4477–4485. doi:10.5194/bg-9-4477-2012

Webb LJ (1978) A general classification of Australian rainforests. *Australian Plants* **9**, 349–363.

Webb LJ, Tracey JG, Williams WT (1976) The value of structural features in tropical forest typology. *Australian Journal of Ecology* **1**, 3–28. doi:10.1111/j.1442-9993.1976.tb01089.x

Wentworth CK (1922) A scale of grade and class terms for clastic sediments. *The Journal of Geology* **30**, 377–392. doi:10.1086/622910

Wetherby K (1984) *The extent and significance of water repellent sands of Eyre Peninsula.* Tech. Report 47. South Australian Department of Agriculture, Adelaide.

White A, Sparrow B, Leitch E, Foulkes J, Flitton R, Lowe AJ (2012) *AUSPLOTS Rangelands Survey Protocols Manual.* University of Adelaide Press, Adelaide, SA.

Wigley B, Charles-Dominique T, Hempson GP, Stevens N, te Beest M, *et al.* (2020) A handbook for the standardised sampling of plant funcitonal traits in disturbance-prone ecosystems, with a focus on open ecosystems. *Australian Journal of Botany* **68**, 473–531.

Williams BG (1988) Subsurface investigations. In *Australian Soil and Land Survey Handbook: Guidelines for Conducting Surveys*. (Eds RH Gunn, JA Beattie, RE Reid, RHM van de Graaff) pp. 154–165. Inkata Press, Melbourne.

Williams J (1983) Soil hydrology. In *Soils: An Australian Viewpoint*. pp. 507–530. Division of Soils, CSIRO. CSIRO, Melbourne and Academic Press, London.

Williams KJ, Harwood TD, Lehmann EA, Ware C, Lyon P, *et al.* (2021) Habitat Condition Assessment System (HCAS version 2.1): enhanced method for mapping habitat condition and change across Australia. CSIRO, Canberra. doi:10.25919/n3c6-7w60

Wood S, Stephens H, Foulkes J, Ebsworth E, Bowman D (2015) *AusPlots Forests Survey Protocols Manual*. Version 1.6. University of Tasmania, Hobart.

Wright MJ (1983) Red-brown hardpans and associated soils in Australia. *Transactions of the Royal Society of South Australia* **107**, 252–254.

APPENDIX 1

Items removed in this edition of the Field Handbook.

Chapter	Item
Preface	Preface to first, second and third editions
Location	Air photo reference
Landform	Riseslope and risecrest landform elements
Land surface	Runon velocity

APPENDIX 2

The peat definition for Australia was developed by the national peat working group and is adapted from a variety of sources including international, national and state definitions.

Peat is defined as:

1. Partially to completely decomposed organic material that is derived from plants, formed under episodic to permanent waterlogged conditions, and;
2. The organic carbon content by dry weight[33] is:
   a. ≥12% if no clay (zero clay %) within the fine earth fraction
   b. ≥13% if the fine earth fraction is ≥10% clay
   c. ≥14% if the fine earth fraction is ≥20% clay
   d. ≥15% if the fine earth fraction is ≥30% clay
   e. ≥16% if the fine earth fraction is ≥40% clay
   f. ≥17% if the fine earth fraction is ≥50% clay
   g. ≥18% if the fine earth fraction is ≥60% clay

Note that the percentage of clay is within the fine earth (mineral) fraction. Peat or peat-like materials generally have a low fine-earth fraction (as a proportion of any given volume).

This definition is such that some materials may meet the criteria of a peaty pseudo-texture, but not meet the technical definition of peat – these are *peat-like* materials. Furthermore, the strict determination of material as peat is reliant upon analytical data, as well as field data. In general terms, peat is formed in waterlogged environments in which the production/ preservation of organic matter exceeds the rate of loss. Peats have a number of characteristic properties, including a very low bulk density compared to mineral soils, and a very high water storage capacity (hundreds to thousands of per cent; Shien *et al.* 2011). Their bulk density is influenced by the degree of decomposition, mineral fraction and carbon content (Warren *et al.* 2012). While peat formation is associated with waterlogged conditions, it is not uncommon in Australia for dry peats to occur – either as a result of seasonal/climatic variability, or due to artificial drainage.

---

33   Walkley-Black x 1.3 or a total combustion method (Rayment and Lyons methods 6A1 or 6B2)

APPENDIX 3

The following are examples of horizon/layer nomenclature designations for polygenetic profiles, lithologic discontinuities, buried soils and fining-up sequences. These examples are by no means exhaustive, but should provide sufficient guidance on the allocation of nomenclature in most scenarios.

POLYGENETIC PROFILES

In polygenetic profiles, more than one set of pedogenetic conditions have occurred in space and/or time, leading to the formation of more than one recognisable profile (or portions thereof) within the vertical section observed. These differing conditions may or may not be related to a lithologic discontinuity/soil burial. Figure 30 provides two examples of polygenetic profiles.

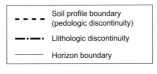

Figure 30   Examples of polygenetic profiles with associated horizon nomenclature (note that texture is not the only attribute used to determine the horizon nomenclature, it is merely given as contextual information).

Profile A shows the development of two sequences of A2 horizons, the lower of which is underlain by a pan. There may be a lithologic discontinuity present, but the evidence is not conclusive. The pan is a critical feature affecting genesis of the lower profile (through ponding of water in occasional periods of flooding or very high rainfall), whereas genesis of the upper profile is governed by surficial factors (leaching to a limited depth due to an arid environment).

In Profile B, a lithologic discontinuity is clearly evident at depth (the 3D horizon). It is influential in the formation of both the upper and lower profiles, and the polygenetic nature in this case is a function of seasonal conditions (wet season *vs* dry season) – that is, two soil profiles have formed within the one material (sand) deposit as a result of varying pedogenetic conditions (a seasonally variable watertable in beach sands overlying marine clay).

In both examples, the 2A2 horizon could be regarded as a 2C (with subsequent adjustment of the nomenclature of underlying horizons) if the observer determined that the horizon lacked any pedological development. In both examples, there is a localised (vertical) separation of modern pedogenetic conditions. In other examples of polygenetic profiles, there may be significant separation in time of pedogenetic conditions, for example, glacial and post-glacial conditions. Given the impact of agriculture (in particular cultivation) on soils, it can be argued that many agricultural soils are now polygenetic.

From the point of change, the underlying profiles are numbered sequentially, downwards, starting at 2. The upper soil is not numbered 1, this being understood.

LITHOLOGIC DISCONTINUITIES

Lithologic discontinuities are **obvious** contrasts in *lithology*[34] between adjacent horizons/layers in the soil profile. They are typically found in alluvial deposits, but can also be found in hillslope environments. Key indicators can be a significant change in proportions of clay, sand or silt content, the size of the sand fraction or lithology of coarse fragments. Field texture is not necessarily a direct measure of lithology, but is typically a strong indicator of changes in lithology. Lithologic discontinuities should be discernible in the field, but laboratory data are often useful in confirming them.

Below the point of lithologic discontinuity, different lithologic units are given a numeric prefix in the manner used for pedologic discontinuities. Each lithologic unit may consist of one or more horizons/layers. Discrimination of lithologic discontinuities from soil horizon boundaries will depend in each case upon evidence of pedological organisation and the contrast between lithologic units. Over time, a lithologic discontinuity may become less obvious as pedogenetic processes operate on the horizons as a whole and mixing occurs through processes such as bioturbation.

Figure 31 provides a number of examples. Profile C is a relatively straightforward scenario in which the lithologic discontinuities are obvious via significant changes in texture. Profile D is an example in which the only indicator of the first lithologic discontinuity is the presence of gravel and the uppermost five horizons (which span three depositional units) have evolved into a singular profile over-printing on the lithologic discontinuities – hence the continuation of the B horizon nomenclature through the depositional layers. Such examples are often difficult to discern.

In Profile E, the lithologic discontinuity is subtle. It is indicated by the presence of appreciable fine sand in the lower horizons and by other factors, not shown here (rapid increase in salinity

---

34   Refer to the Substrate chapter for further discussion of lithology

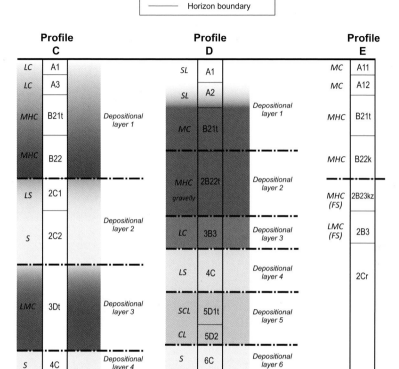

Figure 31 Examples of lithologic discontinuities and a fining-up sequence, with associated horizon nomenclature (note that texture is not the only attribute used to determine the horizon nomenclature, it is merely given as contextual information).

and sodicity), associated with the underlying substrate, which is a different rock type to that which the uppermost four horizons were derived from. These horizons were deposited on the lower horizons through colluvial processes but all of the horizons are undergoing pedogenesis as a profile – hence the continuation of the B2 horizon subdivisions through the lithologic discontinuity. In the strictest sense, the upper horizons have buried the lower horizons (the upper is colluvium, the lower are formed *in situ*); however, the lower horizons are not deemed to be buried, as the whole profile is forming collectively (at the same time) from the two parent materials and the two key processes (aggradation and weathering *in situ*). If it was determined that the lower horizons had formed first, and the upper profile had been deposited over the top in a separate process, then the lower profile would be regarded as buried.

Where lithologic discontinuities are suspected in the profile but there is no clear evidence, either the numeric prefix should not be used or a query should be added to indicate doubt. It is better used only where there is clear evidence. The suffix b, for horizons within buried soil profiles, is not used in the examples in Figure 31. It is also the case in these examples that the D horizons, while having evidence of pedologic development, could not be attributed clearly to a B horizon of a buried soil profile – a pre-requisite for the use of the b suffix (see below).

BURIED SOILS

Buried soils are relict soils that were formed on an older land surface and have been subsequently buried by a younger deposit that in turn has undergone pedogenesis (cf. M layer). Depending on the depth of overlying material, they may or may not be affected by current pedogenic processes. By definition, buried soil profiles involve more than one instance of soil genesis, thus they can be regarded as polygenetic, but this is a differing context to that discussed above and in Figure 30. Recognition of a buried soil relies upon careful observation and determination that a recognisable underlying soil profile is present – that is, soil materials that have undergone pedogenesis as a profile *prior to* being buried.

A buried soil may be truncated – for example, the A1 horizon may have been removed by erosion prior to deposition of the overlying material, or it may have been obliterated by bioturbation post-burial. Where it is possible to *confidently* designate the horizon nomenclature in buried profiles, the buried horizons are given the suffix 'b', which is written last.

A prefix number is used in the same manner as described for other pedologic discontinuities. The same numeric prefix is applied to all horizons/layers within the buried soil (unless there is another pedological discontinuity encountered).

Two examples are given in Figure 32. In Profile G, the nature of the buried profiles is discernible. In Profile H, it is not possible to accurately determine where in a profile the buried horizons lay, although there is sufficient evidence to indicate that pedologic development led

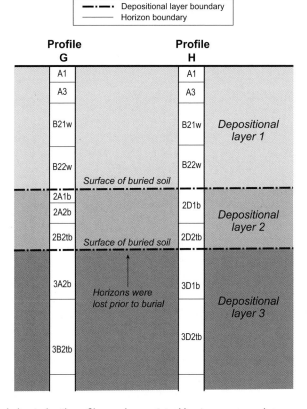

Figure 32 Example buried soil profiles and associated horizon nomenclature

to the development of a profile – that is, it is possible to discern that you are observing part of a historical (buried) soil profile, but you are not sure exactly which horizon it was. This is contrasted with the presence of D horizons in Profiles C and D in Figure 31 in which pedologic development is evident but a b suffix is not applied, because in those examples there was insufficient evidence of a buried profile.

FINING-UP SEQUENCES

A common scenario encountered in alluvial soils is that of a fining-up sequence, depicted as Profile F in Figure 33. In such a situation, clay content may increase with depth to a maximum in the B2 horizon and then decrease with continuing depth. This **does not** constitute a lithologic discontinuity, as it is by definition a continuous feature – typically a result of decreasing fluvial energy over time. The sequence of horizons/layers is, however, likely to be bounded by a lithologic discontinuity. In Profile F, the sequence finishes with a sand C layer, but this is not regarded as a lithologic discontinuity in this instance, as the sand is the commencement of the fining-up sequence.

The nature of horizon boundaries is important when determining if a fining-up sequence is present, as are multiple other factors, and it often requires considerable skill, knowledge and data in order to determine if a sequence of horizons is a real fining-up sequence or involves lithologic discontinuities.

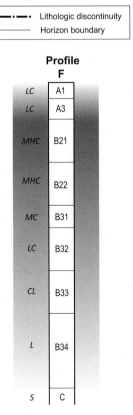

Figure 33 Example of a fining-up sequence with associated horizon nomenclature.

INDEX

Where more than one page reference is given, the numbers in **bold** type indicate reference to definitions or principle discussion.